Management of Convergence in Innovation

Contributions to Management Science

www.springer.com/series/1505

M. Bonilla/T. Casasus/R. Sala (Eds.)
Financial Modelling
2000. ISBN 978-3-7908-1282-4

S. Sulzmaier
Consumer-Oriented Business Design
2001. ISBN 978-3-7908-1366-1

C. Zopounidis (Ed.)
New Trends in Banking Management
2002. ISBN 978-3-7908-1488-0

U. Dorndorf
Project Scheduling with Time Windows
2002. ISBN 978-3-7908-1516-0

B. Rapp/P. Jackson (Eds.)
Organisation and Work Beyond 2000
2003. ISBN 978-3-7908-1528-3

M. Grossmann
Entrepreneurship in Biotechnology
2003. ISBN 978-3-7908-0033-3

H. M. Arnold
Technology Shocks
2003. ISBN 978-3-7908-0051-7

T. Ihde
Dynamic Alliance Auctions
2004. ISBN 978-3-7908-0098-2

J. Windsperger/G. Cliquet/G. Hendrikse/M. Tuunanen (Eds.)
Economics and Management of Franchising Networks
2004. ISBN 978-3-7908-0202-3

K. Jennewein
Intellectual Property Management
2004. ISBN 978-3-7908-0280-1

M. J. Thannhuber
The Intelligent Enterprise
2005. ISBN 978-3-7908-1555-9

C. Clarke
Automotive Production Systems and Standardisation
2005. ISBN 978-3-7908-1578-8

M. Lütke Entrup
Advanced Planning in Fresh Food Industries
2005. ISBN 978-3-7908-1592-4

U. M. Löwer
Interorganisational Standards
2006. ISBN 978-3-7908-1653-2

G. Reepmeyer
Risk-sharing in the Pharmaceutical Industry
2006. ISBN 978-3-7908-1667-9

E. Kasper
Internal Research & Development Markets
2006. ISBN 978-3-7908-1728-7

L. Coleman
Why Managers and Companies Take Risks
2006. ISBN 978-3-7908-1695-2

M. A. Bader
Intellectual Property Management in R&D Collaborations
2006. ISBN 978-3-7908-1702-7

David L. Cahill
Costumer Loyalty in Third Party Logistics Relationships
2007. ISBN 978-3-7908-1903-8

G. Cliquet/G. Hendrikse/M. Tuunanen/J. Windsperger (Eds.)
Economics and Management of Networks
2007. ISBN 978-3-7908-1757-7

Hartmut Hübner
The Communicating Company
2007. ISBN 978-3-7908-1928-1

Jan M. Deepen
Logistics Outsourcing Relationships
2007. ISBN 978-3-7908-1916-8

Eike A. Langenberg
Guanxi and Business Strategy
2007. ISBN 978-3-7908-1955-7

Oliver Falck
Emergence and Survival of New Businesses
2007. ISBN 978-3-7908-1947-2

Christoph Kausch
A Risk-Benefit Perspective on Early Customer Integration
2007. ISBN 978-3-7908-1961-8

Fredrik Hacklin

Management of Convergence in Innovation

Strategies and Capabilities for Value Creation Beyond Blurring Industry Boundaries

Physica-Verlag
A Springer Company

Dr. Fredrik Hacklin
ETH Zurich
Department of Management,
Technology and Economics
Kreuzplatz 5
8032 Zurich
Switzerland
fredrik.hacklin@alumni.ethz.ch

ISBN 978-3-7908-1989-2 e-ISBN 978-3-7908-1990-8

DOI 10.1007/978-3-7908-1990-8

Library of Congress Control Number: 2007936204

Production: LE-TEX Jelonek, Schmidt & Vöckler GbR, Leipzig
Cover-design: WMX Design GmbH, Heidelberg

Printed on acid-free paper

9 8 7 6 5 4 3 2 1

springer.com

Advance Praise

for *"Management of Convergence in Innovation"*

"This book should be read by anyone interested in how firms can combine business models with innovation towards improving their performance. The book is very insightful with a number of eye-opening examples. It is very well grounded in the current realities of the telecommunications industry, and executives in this industry will gain a lot of highly valuable advice for their development of new business models."

—Georg von Krogh, Professor of Strategic Management and Innovation, ETH Zürich

"Much like technological convergence forces industry to reassess itself, this book forces the reader to reconsider our thinking about the roles of industry and technology in innovation."

—Gary Clayton, Vice President Speech Applications, Yahoo! Inc.

"The prospect of convergence represents a highly relevant challenge for managing the pipeline of new technologies. The need for integrated architectures due to shrinking form factors, in combination with the functional fusion of formerly distinct radio technologies, allow us to position ourselves as a platform player in the emergent ecosystem. This work provides deep insights on how to navigate through the process of convergence, and offers instrumental guidance for building such a position."

—Arup Gupta, Director Wireless Platform Technologies, Ultra Mobile Group, Intel Corp.

"In the world of mobile applications, major strategic challenges are rooted in the current convergence trends between IT, telecom, fixed and mobile technologies. Whereas we encounter new opportunities for capturing value through mobilizing traditionally 'inside-the-four-walls' services, we are also increasingly faced by new competitors. For understanding and reacting upon these paradigmatic changes, various analytical frameworks presented in this book provide a comprehensive basis for managerial decision-making."

—Ari Backholm, Vice President Marketing & Product Management, Seven Networks Inc.

"Innovation spawns evolutionary and revolutionary change at an ever-accelerating rate. We can either be surprised, disrupted, and flattened by this change, or we can take a proactive, offensive, forward-leaning approach to anticipate, embrace, and help catalyze innovation driven by technological and business model convergence. This insightful and far-reaching study helps to explain, and advances our collective understanding of, convergence in innovation, so we can meet it head-on, and therefore capitalize on the opportunities it presents."

—Mark Yolton, Vice President, SAP Community Network, SAP Labs, LLC

Preface

Driven by the fascination about dramatic structural and competitive changes within telecommunication and information technology industries during the past decade, the convergence phenomenon has increasingly gained my personal attention throughout my work and studies. Therefore, not entirely coincidentally, this book was written as the result of my doctoral research at ETH Zurich, which turned out to be a challenging, yet highly rewarding endeavor. However, this work would not have been possible without the enduring support of several people.

First, I would like to express my gratitude to my thesis supervisor Prof. Fritz Fahrni, for providing me with the opportunity to conduct exciting research projects in close collaboration with industry, and for supporting me with solid guidance and advice all the way. Also, I would like to thank Dr. Christian Marxt, for urging me to pursue the chosen line of enquiry, as well as for his devoted coaching, both at ETH and at Stanford, both within and beyond office hours. Furthermore, I am grateful to Prof. Georg von Krogh, for his encouraging feedback and valuable comments during various inspiring discussions.

For the good collaboration within an excellent working atmosphere, I would like to thank the entire team of the Chair for Technology Management and Entrepreneurship, as well as various colleagues, alumnis and friends from ETH Zurich, especially Dr. Nešo Atanasoski, Andrin Blauenstein, Avrath Chadha, Beat Gyger, Monica Heinz, Martin Inganäs and Dr. Vicente Raurich. Also, I am indebted to all my students who contributed to this research within various projects and master's theses during the past years, particularly, thanks to Julian Sommer for faithfully helping me out in the never-ending data transcription. For excellent librarian support I am grateful to Dr. Hanspeter Schwarz. Special thanks go to Dr. Oliver Blauenstein and Dr. Jan Malmström, who brought me in touch with this captivating institution in the first place.

At Stanford University, I would like to express my gratitude to the Scandinavian Consortium for Organizational Research (Scancor), particularly to Prof. Woody Powell, Prof. Jim March and Barbara Beuche. For contributing to the progress of my research through various fruitful insights and impulses, I am grateful to Prof. Antti Ainamo, Sam Garg, Dr. Jason Davis, Prof. Riitta Katila, Prof. Larry Leifer, Dr. Matti Perttula, Sari Stenfors and Dr. Martin Wallin. In this context, I would also like to acknowledge the Scancor seminar on organizational science, the Scancor informal seminar, the Monday "munch" seminar, as well as the DesignX wine and cheese meeting.

I am particularly indebted to all participating case companies, who were eager to participate in the study, and invested a remarkable amount of time for various interviews and discussions. Special thanks go to Nina Granqvist, who was a highly reliable partner during numerous hours of data collection. Also, I would especially like to thank Dr. Judy Kleinberg for her interest and contribution to this research.

For exchanging various ideas and research results related to the field, I am grateful to Dr. Niklas Adamsson, Dr. Jukka-Pekka Bergman, Davide Chiaroni, Jan Edelmann, Jouni Koivuniemi, Anna-Greta Nyström and Prof. Yves Pigneur. For taking over the burden of proofreading the manuscript, special thanks go to Morgan Reino Altman.

For facilitating my research stay at Stanford, I am indebted to the Helsingin Sanomat Centennial Foundation as well as the Huber-Kudlich Foundation.

Finally, I owe my biggest thanks to my parents and Aino, who always have supported and encouraged me throughout my studies.

Zurich, May 2007 *Fredrik Hacklin*

Contents Overview

Table of Contents

List of Figures

List of Tables

1

Introduction

> "Transport of the mails, transport of the human voice, transport of flickering pictures—in this century as in others our highest accomplishments still have the single aim of bringing men together."
>
> *Antoine de Saint-Exupéry (1900–1944)*

Not so long ago, telephones and computers had relatively little in common. Whereas phone call switching and dialing were based on analog signal transmission, computing technology had its roots in storing and processing digital data. Both technologies had their own inventors, their own set of firms in charge of advancing the development, their own industrial worlds, with their own competitive rules. Whereas the one world generated income on minutes, the other did equally so by bytes.

Then there appeared the Internet. Although both the telephone and the computer world embraced the opportunity as a promising application for their respective industries, many involved firms failed to see that something started to happen *between* both worlds, rather than *within* each of them. In the years of Internet pioneering, the consumer was able to use his analog phone line for connecting to the World Wide Web. Technically, a packet-switched communications protocol was used on top of the circuit-switched phone line. With the evolution of broadband networks, however, the shape of Internet access became inverted. Today, many customers have packet-switched Internet access independent of any phone subscription, and can in fact use this technology for establishing Internet-based phone calls. As a result, both initially separate worlds, with their distinct sets of industry players, are today competing for the same users.

When innovations emerge at the intersection of established and clearly defined industry boundaries, they do not only allow for novel applications and improved customer experience in both worlds. The combination, confluence and convergence of technological trajectories may yield outcomes which in their performance exceed the sum of their parts. As a consequence, established paradigms will be replaced by new ones, which introduce new ways of solving problems that were already thought of as solved, and thereby disrupt and substitute rules of conducting business.

In other words, whereas the phone line at some stage was used for getting into the Internet, the Internet can today be used for getting onto the phone. The peculiarity of such industrial turning points, the need for a deeper understanding of this phenomenon, and the question about reproducibility of this form of change represent the underlying rationale of this work.

1.1 Background and motivation

1.1.1 Practical background and relevance

Whereas the concept of *convergence* has represented a subject of attention within the telecommunications and information technology (IT) sectors for over a decade now, recent industrial trends suggest that the initially developed visions finally start to become operational, and that the convergence phenomenon increasingly is jumping "from drawing board concept into consumers' hands, [...] bringing users' media consumption closer to the nirvana of anything, anytime, anywhere" (Accenture, 2006, p. 3). Recognizing the pervasiveness of convergence around digital technologies, the world's largest annual trade show for information and telecommunications technology, CeBIT, was arranged in 2005 under the catch-phrase "digital convergence". Similarly, in a 2004 press release, the telecommunications handset and equipment manufacturer Nokia explained how historically separate industries, i.e., the communications, consumer electronics, and media industries are increasingly coming together into a single industry: "the converged digital industry", which for consumers results in attractive products and solutions, and from a business perspective creates opportunities in new untapped areas (Nokia, 2004, p. 1).

However, the promise of convergence is not only associated with opportunities and growth. As a consequence of the technological coming-together of both worlds, a variety of industrial consolidation activities have taken place, e.g., through mergers or acquisitions

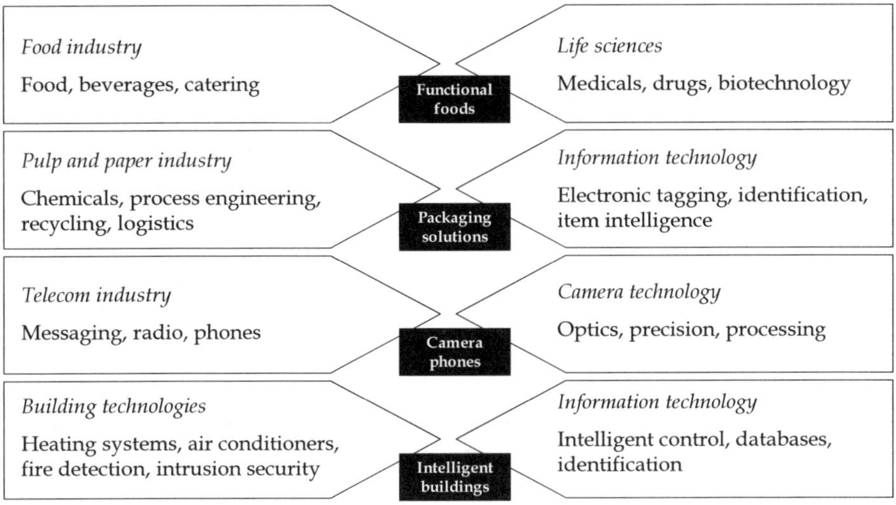

Fig. 1.1: Examples for convergent developments between industries

(Chan-Olmsted, 1998; Lee, 2003). On the other hand, as business models tend to collide (Hackler and Jopling, 2003), convergence may even shake players entirely out of business (Shepard, 2002; Steinbock, 2003).

Additionally, the observation of firms in the context of the pre-study (section 1.4.3) yielded an initial understanding on perceived managerial challenges related to the convergence phenomenon. In particular, managers of technologies in converging environments experience challenges in determining and articulating the implications for strategy development, as well as deriving and operationalizing implied tasks within a structured framework. Within various organizations, several internal activities of local strategizing and restructuring take place, in response to the insecureness regarding the severeness of the changing business landscape. A lack of a more deeply-rooted and grounded understanding of the phenomenon could be observed, which however represents a prerequisite for deriving such managerial guidelines. In many cases, managers felt insecure on how to find a balance between action and reaction. In particular, this balance consisted of positioning strategic intent of search and innovation, on the one hand, thereby actively shaping the renewal, and on the other hand, passively reacting within an environment of emerging new industry structures, with their strategic implications.

After all, the phenomenon may not solely be limited to telephone and computing-related technologies. Figure 1.1 depicts examples for convergent development trends between industries. In particular, emerging scientific advances in the intersection of microelectronics, material design, molecular biology, as well as complex chemistry result in nanoscale developments, which may trigger similar industry phenomena. The consulting firm McKinsey & Company estimates that the cumulative market for converging info-, bio-, and nanotechnologies could top $1 trillion within a decade (SVEDA, 2005).

1.1.2 Shortcomings of existing research

Steinmueller (2000) argues that after two decades of movement towards convergence, the result is still ambiguous. In his view, this might be due to the *ex ante* definition of convergence on a theory level, which however represented an idealization that was necessary to explain the implications of market developments opened by technological opportunity. Whereas there exists a broad literature base focusing on rather technological developments and business models opportunities rising from convergence,[1] the association between convergence and innovation management can be regarded as a rather rarely identified subject.[2] In many cases, current approaches seem to rather focus on symptoms—dealing with the new situation within the firm—than on examining the nature of the underlying mechanism for a rather proactive, participative approach to dealing with convergence. Additional exploratory and explanatory research is needed in order to understand convergence developments as a form of innovation, and in particular, to further relate the dynamics of convergent systems to respective dynamics of innovation processes. Pennings and Puranam (2001) conclude that all convergence processes have an impact of eroding boundaries between industries, which in turn poses challenges to firms, and forces them to face new technologies, consumers and needs. Also, Collis, Bane, and Bradley (1997) observe that industry convergence has inspired radical responses in corporations, but many of those moves fail to be explained with existing frameworks. The emergence of such types of—disrupting and breakthrough—effects of the phenomenon gives rise to the question whether convergence should

---

[1] Cf. Bannister, Mather, and Coope (2004); Fusaro (2002); Parodi and Liggieri (2003); Sigurdson and Ericsson (2003); Theilen (2004)

[2] Cf. Dowling, Lechner, and Thielmann (1998); Harianto and Pennings (1994); Lei (2000)

be approached from a more innovation theory-oriented perspective, in order to result in more comprehensive frameworks. Moreover, it is argued that the convergence phenomenon should take a center stage in the research and theory on technological change, innovation and corporate strategy (Pennings and Puranam, 2001).

Based on the networked and exogenous nature of the convergence phenomenon, several scholars have previously indicated that inter-organizational dynamics, such as collaborative innovation, strategic partnerships or alliances, represent a major attribute of action in such environments.[3] As technological convergence may infringe on current value propositions of firms, the phenomenon may require firms to take sudden operative responses before careful planning has been made. Hence, as organizational roadmaps for the strategic positioning within an existing ecosystem may be punctuated through emerging impacts of the convergence phenomenon, existing strategic approaches need to be quickly adapted into the emerging paradigm. Hence, the strategic challenge of convergence can be related to an implied need for opening-up existing structures and processes, facing the external phenomenon of convergence through an internal response of divergence.[4]

Whereas previous research suggests that convergence implies collaboration not only beyond firm, but even industry boundaries, current theory falls short in further characterizing this link and providing guidelines. In other words, whereas a relationship between convergence and collaboration is being perceived as intrinsic, it remains rather ambiguous what this association implies in terms of strategic management, i.e., where and when actions should be taken and where not. Hence, the characteristics of the intersection between convergence and inter-firm collaboration have to be further explored.

In his considerations on the "paths to convergence", Steinmueller (2000) would like to see questions answered about how convergence is influencing our lives and industry, aiming at identifying the directions of convergence. In particular, Baer (2004) sees a research need in

---

[3] Cf. Alkemade (2003); Amesse, Latour, Rebolledo, and Séguin-Dulude (2004); Athreye and Keeble (2000); Baer (2004); Blomqvist, Hara, Koivuniemi, and Äijö (2003); Duysters and de Man (2003); Gambardella and Torrisi (1998); Greenstein and Khanna (1997); Gulati, Nohria, and Zaheer (2000); Harianto and Pennings (1994); Hawkins (1999); Lindmark, Andersson, Bohlin, and Johansson (2004); Möller and Svahn (2003); Möller, Rajala, and Svahn (2005); Nyström (2004); Prahalad and Ramaswamy (2000); Rockenhäuser (1999); Varis, Virolainen, and Puumalainen (2004).

[4] Cf. Baer (2004); Götte (2003); Hacklin, Marxt, and Inganäs (2005d); Marxt and Hacklin (2004)

understanding the relationship between co-operations between firms in closely related or neighboring industries, as well as the implications on the value chain in the form of horizontal or vertical integration. The selection of such competitive directions, i.e., the dichotomy between integration along the value chain, or within one single stage of it, can be regarded as sensible from a market dominance perspective, and management tasks arising therefrom should not be underestimated (Wirtz, 1999). In this context, however, there exist to some extent even contradictory conceptions on convergence effects, as on the one hand, convergence may lead to an explosive growth of new market-entrants, on the other hand, however, scholars argue that convergence consolidates the market. In other words, the phenomenon constitutes a major impetus for innovation and economic growth, while at the same time also causing market disequilibrium and firm mortality (Pennings and Puranam, 2001). In particular, a variety of literature contributions indicate that vertical disintegration and value chain deconstruction are phenomena associated to convergence.[5] On the other hand, the true opposite characteristic has been reported in literature as well, i.e., arguing that convergence is promoting a trend towards consolidation, vertically integrated companies and alliances.[6]

This inconsistency of reasoning on the one hand, the consistency of *de facto* motivations for both types of effects on the other, gives rise to the question whether this observed behavior might change over time. In other words, further investigation is needed for understanding to what extent both observations actually coexist and whether the trajectory of disintegration and integration follows a certain sequential and temporal pattern.

---

[5] Cf. Brusoni and Pavitt (2003); Kim (1999); Li and Whalley (2002); Mueller (1999); Pavitt (2002, 2004); Rosenberg (1963); Stigler (1951); Zerdick (2000)

[6] Cf. Bower (2001); Gaines (1998); Henten (2004); Lang (2003); Ramos, Feijoo, Perez, Castejon, and Segura (2002); Wirtz (1999); Yoffie (1996)

An illustrating example for this contradiction can be made by contrasting two quotations from Brusoni and Pavitt (2003) as well as Lang (2003):

"Opportunities have also emerged for further vertical disintegration between product design and manufacture, based on further technological convergence (i.e., in convergence, based on technical change, in specific production operations across firms, products or industries)." (Brusoni and Pavitt, 2003, p. 4)

"One of the most important dimensions of convergence is vertical integration, i.e., the merging of formerly independent firms along the value chain." (Lang, 2003, p. 2)

1.2 Research problem

1.2.1 The quest for an evolutionary understanding

As a consequence, the question arises to which extent these exogenous dynamics can be associated with sequential phases of convergence, and whether there exists such a thing as a *convergence process*. Going back to the generic relationship between convergence and innovation processes, further elaboration on antecedents and implications from an innovation theory perspective are needed in order to understand the dynamics of convergence over time as well as its relation to the evolution of technological trajectories. In this context, Lind (2005) suggests the investigation of convergence by using the technological life-cycle perspective as an underlying approach, relating the phenomenon to existing concepts such as product architecture and dominant design (Abernathy and Utterback, 1978; Utterback, 1994). In this context, also the background, size and history of involved firms, i.e., entrants or incumbents, is identified as a challenging issue in terms of trajectory modeling (Yoffie, 1996).

Additionally, this interplay of cause and effect of convergence requires a bidirectional perspective, as in current literature, focus is laid on understanding firms' reactions to different kinds of convergence. The vice versa, i.e., the influence of firms' actions on the nature and location of convergence, is suggested as highly relevant in this context, as well as the issue of how individual firms shape the boundaries of their industries (Greenstein and Khanna, 1997). Once industries boundaries are shifted, the scope of the dynamics of innovation is altered. Hence the effect of convergence on innovation typology also deserves further attention within the examination of the process.

Gaining an understanding of the convergence phenomenon from such a rather sequential and chronological perspective may provide a basis for deriving firm strategy-specific practices as well as operationalizing convergence-oriented activities into existing organizational processes. The above introduced convolution, resulting from an innovation process based perspective on convergence, i.e., the issue of convergence processes, has been mentioned in earlier literature, however without any explicit definitions (cf. Cockburn, Henderson, and Stern, 2000; Duysters and de Man, 2003; Grant, 2002; Stieglitz, 2003; Wirtz, 1999).[7] In the call for such a sequential perspective, Yoffie (1996)

[7] Particularly, in the existing literature base it remains unclear whether the notion of a "convergence process" refers to the economic change over time from an in-

already notes that one major problem can be seen in predicting the appropriate time frame as well as the course of convergence. Also, Steinmueller (2000) identifies open questions in how rapidly convergence is gaining influence, i.e., on the rate of convergence. Baer (2004) suggests that the convergence development in further research should be studied over time, and establishes a link between convergence processes and inter-firm collaboration. In particular, while his work primarily focuses on the strategic decision to enter co-operations, the partner selection as well as the configuration of cooperative arrangements within converging environments, this opens the question for convergence-related implications on such cooperative activities and portfolios over time (Baer, 2004). In the work of Pennings and Puranam (2001), a variety of possible means for measuring convergence through observing certain inter-organizational activities is listed and the authors suggest these to be monitored over time in order to expose trends.[8]

1.2.2 Towards a capability-oriented perspective

Finally, in converging systems, innovative approaches need to be pursued in order to gain and sustain competitive edge in the increasingly heterogeneous and fluctuating environment (Kaluza, Blecker, and Bischof, 1998). While the aspect of collaboration and networks is perceived as an intrinsic element of any convergence related managerial activities, research suggests addressing the phenomenon on a broader and more generic conceptual level. Rockenhäuser (1999) argues that it is not sufficient, to solely develop an understanding

novation theory perspective, or if it should represent a set of operational practices for firms acting in converging environments. Rockenhäuser (1999) and Moschella (1997) attempt to define phases of a convergence process, however remaining on a rather rudimentary level.

[8] These recommendations include the measurement of network based methods relying on inter-firm development activities, i.e., joint ventures, minority interests, licensing, as well as R&D diffusion, i.e., the mobility of scientific and technical personnel as interesting avenues for documenting convergence. Moreover, it is suggested to quantify the creation of consortia spanning communities of practice across industry sectors, as well as the citation of patents belonging to different classes. In this context, the analysis of pre-alliance technical networks is also proposed, referring to the concept of "cooperative technical organizations" as discussed by Rosenkopf, Metiu, and George (2001). Furthermore, also network and graph theory related dimensions brought into the context of convergence, including adjacency indices such as degree centrality, betweenness, density and clique membership (Andrews, 1995; Burt, 1992, 1997; Krackhardt, 1995; Salancik, 1995) are suggested to be examined over time (Pennings and Puranam, 2001).

of the dynamics of industrial value creation processes along a deconstructed and converging value chain. Furthermore, Rockenhäuser (1999) sees the challenge in a successful focus on entrepreneurial competence portfolios based on an underlying heterogeneous set of firms, representing key focal points of a value-oriented convergence strategy, since competencies and core competencies represent the basis for value creation and competitive advantage of a firm. This perspective of managing competencies rather than collaborations formulates a need for further research on the requirements of convergent systems on capability development, i.e., a broader, more generic perception than solely planning collaborative activities.

Given such a perspective, the concept of dynamic capabilities (Eisenhardt and Martin, 2000; Teece, Pisano, and Shuen, 1997) can serve as an integrative model for developing management guidelines in a converging environment. As such an environment can be characterized by change and uncertainty, managerial responses have to focus on developing mechanisms for strategic adaptability and organizational learning, maintaining a continuous window on renewal, in order to sustain competitive advantage.[9] In the work of Lei (2000), the need for certain dynamic capabilities for managing convergence is further identified. Along with the trajectory of industries becoming closer, firms need to invest in new types of core competences and resources, that allow them to deploy value-creating skills and to be easily reconfigured and adapted to serve a wider base of customers across different markets. This poses important strategic and organizational challenges and even existential dilemmas to firms whose business units compete in industries that face shifting boundaries as new approaches to acquiring and internalizing knowledge and skills are required. Hence, there is a necessity for a 'venture capital'-like approach to searching and investing in new technologies, as well as for an organizational design that can support firms in quickly learning multiple types of core competences (Lee, 2003; Lei, 2000).

By employing the concept of dynamic capabilities as an underlying rationale for developing convergence guidelines, the outcome of such guidelines should not be limited to decision and selection support on whether to collaborate or not, with whom and to what extent. Whereas the capability of a value-oriented management of competence gaps

---

[9] One can distinguish between ordinary or zero-level capabilities, being those that permit a firm to 'make a living' in the short term, as well as dynamic capabilities, being those that operate to extend, modify or create ordinary capabilities (Winter, 2003).

is considered as a crucial element of a convergence-oriented strategy (Rockenhäuser, 1999), it shall be avoided to directly associate capabilities as an equal to competencies, but rather as a superset of them. Moreover, it shall be emphasized that this concept leaves room for much more complex constellations of such capabilities. For instance, Möller and Svahn (2003) as well as Möller *et al.* (2005) discuss the concept of network orchestration capabilities as an actor's capacity to influence the evolution of a new value network, and suggest this set of capabilities to be developed for envisaging very complex emerging business fields such as they occur in converging environments, and for identifying potential trajectories therein. As another perspective to capability development, Prahalad (1998) states that a large number of similar convergences can be identified, and regards the major common challenge in managing and seamlessly integrating very distinct intellectual heritages that stem from the variety of industrial backgrounds, histories and associated cultures. Similarly, Pennings and Puranam (2001) see the main challenge among multiple forms of convergence in the effect of degrading or rendering obsolete current organizational capabilities. What in their view makes the identification and development of new capabilities so difficult, however, is the fact that the ability of firms to refashion their capabilities to new demands is inhibited by path dependence (Pennings and Puranam, 2001; Teece *et al.*, 1997). Thus, the implied combination of convergence and path dependence produces a unique tension, as a firm on the one hand is expected to remain coherent with its industry and 'stick to the knitting', while on the other hand convergence induces it to diversify by venturing into new, undiscovered markets. This tension between abiding to the coherence imperative (Teece, Rumelt, Dosi, and Winter, 1994) and acting in a converging environment, requires an elasticity of successfully balancing between specialization and despecialization, and hence, poses a major impetus for research and theory on convergence (Pennings and Puranam, 2001).

Actors in converging environments are imposed by a strategic stretch, which formulates a need to search for an optimal trade-off among various firm strategies. This represents a rather unusual perspective on innovation management practices and stands in contrast to current paradigms.

1.2.3 Industry-independent comprehension

Having already observed the phenomenon of technological convergence in the machine tool industry during the industrialization era,

Rosenberg (1963) suggests the investigation of convergences in other industries as a verification for the theory:[10]

> "A question of more contemporary interest is whether similar technological convergences are occuring in twentieth-century conditions; whether, for example, the chemicals and electronics industries are playing the same roles of information production and transmittal that machine tools played at an earlier stage in our history." (Rosenberg, 1963, p. 443)

More recently, technological convergence has been broadly observed in the ICT industry,[11] and the question of generalizability of findings and transferability onto other industries has emerged among these various studies. Also, Baer (2004) infers that even if solely considering the ICT-driven technological convergence, its implications might possibly reach out further from the industry sectors media, IT and telecom, and could induce development processes within other industries as well. Providing a perspective for further examining technological convergence, Lei (2000) notes that the phenomenon is especially salient to those firms competing in fast-changing, high-velocity environments, as a high degree of intra-firm and inter-firm coordination to learn and build new sources of knowledge, skills and insights is required for sustaining a competitive advantage (cf. Dosi, 1982; Eisenhardt, 1989b; Helfat, 1997; Nonaka, 1991, 1994; Teece *et al.*, 1997). Convergence of technologies has been studied either during the growth hype, or during phases of consolidation, but an integrated, comprehensive view to the convergence process, which could be used for understanding and managing through similar innovation trajectories in future, can be regarded as still missing. Particularly, retrospective considerations, which can be made in environments where the convergence development has reached a certain maturity (such as e.g., ICT), might serve as a basis for being transferred onto other, new, emerging convergences, which may profit from a capability-oriented anticipation and management model.

---

[10] See also section 2.1.2

[11] Cf. Andersson and Molleryd (1997); Bannister *et al.* (2004); Bohlin (2000); Borés, Saurina, and Torres (2003); Duysters and Hagedoorn (1998); Edwards (1999); Fransman (2000); Friedman and Waldman (1992); Joseph (1993); Lee (2003); Nikolaou, Vaxevanakis, Maniatis, Venieris, and Zervos (2002); Rao (1999); Sabat (2002); Sherif (1998); Stieglitz (2002, 2003, 2004); Yoffie (1996)

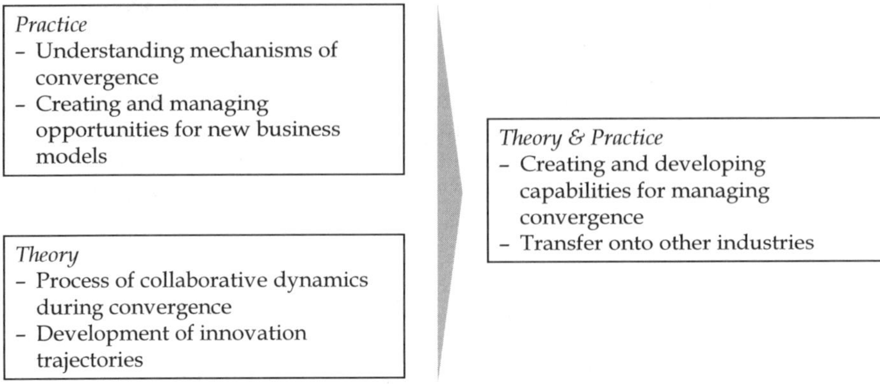

Fig. 1.2: Observed gap between theory and practice

1.3 Research objective

1.3.1 Formulation of research questions

Based on the combination of the above-mentioned challenges from practice, with the identified shortcomings in existing theory (figure 1.2, cf. Ulrich, 2001), the aim of this research is to contribute to a deepened understanding on opportunities for deliberate managerial action taking, in order to influence the dynamics of innovation under the premise of convergence. Hence, the main research question is formulated as follows:

Question 1. (Q_1) *Are there key capabilities which enhance the management through processes of technological convergence?*

Recognizing the fact that the effect of convergence represents an underlying premise, and may not be directly influenceable through deliberate action, it is therefore chosen to regard the phenomenon as an independent variable. As the main question Q_1 aims at understanding the influence of managerial action-taking onto the impetus on innovation trajectories, the dynamics of innovation therefore represent the dependent variable. As the managerial capability is interacting between the phenomenon *per se* and the implied dynamics, it is considered as a variable with moderating effect. Hence, in order to be able to respond to the main question Q_1, the underlying premises yield the following two subquestions:

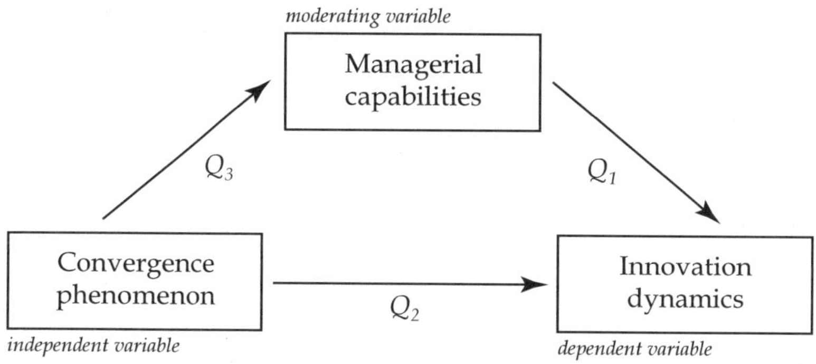

Fig. 1.3: Taxonomy of research variables and questions

Question 2. (Q_2) *How can the dynamics of convergence be comprehended from an evolutionary perspective?*

In particular, question Q_2 aims at understanding to what extent an understanding of the convergence effect of innovation processes and dynamics can serve as a basis for structuring endogenous and exogenous innovation activities. Finally, establishing a link between independent and moderating variables, the following question is posed:

Question 3. (Q_3) *How does technological convergence influence innovation management practice?*

In other words, question Q_3 aims at investigating the effect of the convergence phenomenon on existing capabilities of managerial action. Figure 1.3 depicts a taxonomic framework of research variables, associated research questions, and their mutual interdependencies.

1.3.2 Derivation of research goals

Based on the previously formulated research questions Q_1 to Q_3, an operationalization into goals for this research is made. In particular, four main research goals are derived and summarized in table 1.1, which represent the basis for the structuring of this work (section 1.5).

Table 1.1: Research goals

Identify fundamentals. Literature survey on existing theory on the intersection of convergence and innovation management *(chapter 2)*

- Understanding of current research state of the art
- Causalities of the convergence phenomenon
- Effects of convergence phenomena on innovation systems

Understand process of change. Retrospective investigation of convergence phenomena from an evolutionary perspective *(chapter 3)*

- Links between prior distinct observations during growth hype and consolidation
- Response to the existing lack of a retrospective view on convergence processes
- Extension of theory on dynamics and trajectories through a conceptual anticipative process for innovation management
- Theoretical basis for the translation into managerial processes

Managerial action within process. Investigation of capabilities needed for managing through different convergence constellations with different firm-specific properties *(chapter 4)*

- Extension of understanding of firms' implied reactions to different configurations of convergence
- Identification of possibilities for firms' deliberate actions to different configurations of convergence
- Understanding of influence of firms' actions on the nature and location of convergence

Managerial guidelines and transfer. Consideration and closer examination of applicability onto other, new and emerging converging areas *(chapter 5)*

- Formulation of managerial guidelines
- Recommendations for capability development
- Considerations of emerging technological trends

1.4 Research design

1.4.1 Methodological approach

In Aristotelian logic, the method of induction describes the "argument from the particular to the universal", which represents the basis of knowledge on indemonstrable first principles (Smith, 2004). Given the lack of existing theoretical frameworks for addressing the research questions, an approach of building new, emergent theory, rather than testing initial hypotheses is needed (cf. Ulrich, 2001).

Therefore, a qualitative, grounded theory-building approach is selected (Eisenhardt, 1989a; Eisenhardt and Graebner, 2007; Glaser and Strauss, 1967), allowing for a research design to be located closer to the phenomenon than to predefined theoretical frameworks. Such an approach is feasible in times when little is known about a phenomenon, and current perspectives seem inadequate due to little empirical substantiation, or if existing views conflict with each other or common sense (Eisenhardt, 1989a). With respect to the context of convergence, this approach can be motivated by the fact that the phenomenon from a holistic innovation management perspective still can be regarded as a rather new and diffuse field of research, with only few underlying theory models existing, and finally, with observed contradictory perceptions in literature.[12] Also, the approach of building theory from case study research is regarded as appropriate in early stages of a research area, or in order to provide an already researched topic with freshness in perspectives (Eisenhardt, 1989a).

Drawing on this insight, the selected research design is an inductive, multiple-case study involving 26, respectively 9 firms within two distinct case samples. Through using multiple cases, a replication logic is enabled, in which cases are treated as a series of independent experiments that confirm or disconfirm emerging conceptual insights (Yin, 1994). By doing so, overall cross-case analysis approaches are facilitated (Yin, 1994, 2003), allowing the extensive use of tables, charts and other cell designs to compare several categories at once (Miles and Huberman, 1984). Hence, results are typically better grounded than those of single-case studies, and the external validity of the findings is increased. In particular, the multiple-case approach yields on the one hand literal replication, by replicating similar results at similar conditions, as well as theoretical replication, by predicting contrasting results for predictable reasons in altered conditions (Yin, 1994, 2003).

Since the convergence phenomenon takes place both within firms, as well as between them, holistic approaches with single units of analysis would satisfy the conditions to only a limited extent.[13] Additionally, as the research questions do not focus on examining momentary

---

[12] In particular, the observation of tensions and paradoxes in prior research can be regarded as a starting point for the building of new theory, as uncovering contradictions may move inquiry into new directions (van de Ven and Poole, 1989).

[13] In their suggestions on new directions for theory and research, Pennings and Puranam (2001) argue that studying the phenomenon of convergence presents several empirical challenges. In particular, the challenge consists of the discrepancy between a needed demarcation of industry or market boundaries, which for purposes of measurement need to be in an equilibrium situation. However, the nature

phenomena or discrete events, but ask for a more differentiated study of a continuous phenomenon with causally linked events over time, the required methodology needs to allow for a processual perspective. As processes by nature are embedded in their context, Pettigrew (1997) suggests that the study of firm-level change should be linked to both higher and lower levels of analysis. Particularly, process phenomena *per se* may be embedded into several levels of abstraction (Pettigrew, 1997), which gives rise to multiple lenses at multiple levels of analysis. Given the multitude of relevant layers of abstraction when dealing with convergence from an innovation management perspective (table 1.2), an embedded design with multiple levels of analysis is therefore applied as an additional configuration of the multiple-case approach, allowing a richer and more reliable model to be induced (Klein, Tosi, and Cannella, 1999; Yin, 2003). The main unit of analysis is the sample of the focal firms within the context of convergence, and the higher levels of analysis consist of mutual interactions between firms, as well as respective industry landscapes. The need for the level of mutual interactions became especially relevant during data collection (section 1.4.3), as single firms serendipitously started to make reference to each-other within the sample.

In the cross-case comparison, the emergence of similar themes and constructs across multiple cases and multiple lenses was looked for. From the patterns that emerged through this procedure, tentative propositions were formed. Based on the replication logic, initial propositions were refined, through continually revisiting data to systematically compare and verify the occurrence of specific themes within separate cases. In contrast to theory-testing approaches, the coding of data does not represent a prerequisite for analysis. In a grounded theory-building approach, a theoretical framework does not exist yet and is subject to emerge from the research process. In such a setting, the act of coding represents the process of concept development, containing an iterative discussion between empirical material and induced theory (Glaser and Strauss, 1967; Strübing, 2004). Through such a continuous iteration between theory and data, constructs could be sharpened, and the internal validity of findings could be strengthened (Yin, 2003).

of convergence is by nature a disequilibriating process, raising a need for multiple research objects as well as units of investigation.

Table 1.2: Abstractional embeddedness of research questions

Research question	*Level of analysis*
Q_1	firm
Q_2	industry
Q_3	firm/industry

1.4.2 Research frame and setting

According to Kubicek (1975), a multilevel research design allows the integration of both contextual analyses as well as reductionistic approaches within single enclosing frames of reference, which may contribute to overcoming previously existing one-sidedness. Based on further assigning attributes to the research variables as addressed by the research questions (figure 1.3), and encapsulating the structure of interdependencies into a wider context, the frame of reference for this study is specified. In particular, aspects regarding the antecedents of the phenomenon, albeit only indirectly associated to the research questions, are included in order to understand the phenomenon in greater detail (figure 1.4).

The general research setting is in the context of convergence between industries, with focus on firms active in such intersectional en-

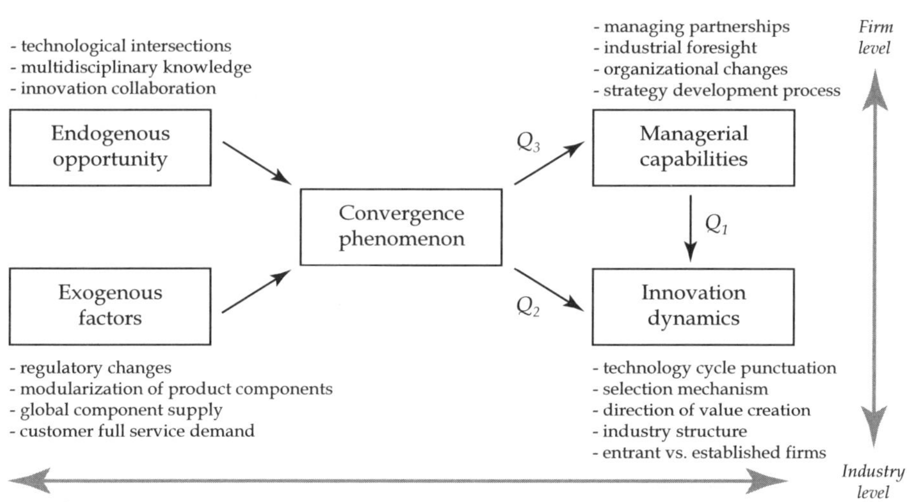

Fig. 1.4: Frame of reference

vironments. For the first firm sample, the setting consists of the scope of information and communication technology (ICT) industries (table A.1, p. 210). This setting was chosen after extensive exploratory work in examining examples for technological convergence within the pre-study (section 1.4.3). The ICT industry represents such a vivid example for convergence, as current and emerging technological disruptions can be characterized based on the existence of economies of scale evoked through digitalization, a high degree of network effects and the relevance of critical mass effects (Baer, 2004). In their confluence, these disruptions evoked remarkable implications on the competitive constellation within underlying original industries. For the second firm sample, the industry setting is based on emerging convergence trends around scientific advances in nanotechnology (table A.2, p. 212). In particular, the second firm sample is aimed at finding similarities and achieving theoretical replication onto emerging convergence developments, based on the model as induced from the study of the ICT set. The latter set was selected, as emerging trends of interdisciplinary product development and structural industry changes in the context of nanotechnology were observed within literature and practice (section 5.3.2).

Due to the nature of replication logic in contrast to the sampling logic, all firms within the sample were chosen because they were claimed to have positive outcomes in terms of technological convergence beforehand. The case studies and the ensuing evaluation then predict that similar processes would be found to account for these outcomes (i.e., direct replications). If such replications are indeed found for several cases, a higher level of confidence in the overall result can be achieved. The development of consistent findings, over multiple cases or even multiple studies, can then be considered a more robust finding within the given contextual setting (Yin, 1994, 2003).

The selection of the case sample was based on an initial screening process, aiming at investigating whether a respective firms fulfill the conditions of being affected by converging technologies and increasing interdisciplinarity in product development practices. This screening process occurred prior to the actual conduct of the case studies. If a subsequent case study did not turn out to entirely corroborate with the initial screening, the discussion related to a respective firm's positioning represented a substantial part when classifying the cases. Based on such a classification, the resulting cross-case study was able to serendipitously take advantage of the inadvertent outcomes of the screening, by complementing the originally intended direct replications through a reflection of both direct and theoretical replications (Yin, 1994, 2003). The individual case studies attempt to elaborate on convergence patterns, processes, and managerial action within single firms. The purpose of the cross-case analysis is to determine whether the single observations of firms share more generic, common patterns, aiming at identifying similarities, differences, and common themes on an industry-level.

The geographic scope covers Europe and the USA. The organizational scope contains both private and publicly owned high technology firms. In terms of organizational size, the study distinguishes between entrant firms, as well as established firms (section 3.3.1).

1.4.3 Research process

In case-study research, the process shall on the one hand be directed toward the development of theory which is generalizable across settings, allowing generation of further testable constructs (Eisenhardt, 1989a; Eisenhardt and Graebner, 2007). On the other hand, allowing applicability and external validity, both beginning and termination of the process need to be rooted in a practical context (Ulrich, 2001). Building on these guidelines, the research process of this study consists of three main structural phases (figure 1.5).

Following a first identification of the research area, phase 1 was conducted during the years 2003 and 2004. The aim of this pre-study was the formulation of the specific research questions based on an initial field study. Data collection took place in two fields. On the one hand, insights on current and emerging managerial challenges regarding convergence and innovation were developed and refined during the attendance of conferences, workshops and the management meeting of a wireless handset manufacturer.[14] On the other hand, data was collected during the elaboration of distinct case studies within supervised student projects and master's theses in collaboration with partner firms in the ICT industry.[15] During the participation in specific corporate development projects during these collaborations, informal interactions and firm-internal archival data served as main sources of primary, respectively, secondary data. These projects were treated as holistic, single case studies (Yin, 1994), as they apart from exceptions, were conducted in distinct contextual settings. In parallel to these field

[14] Management meeting, firm δ_2 (cf. table A.1, p. 210), 18. December 2003.

[15] Partial results of these projects are reported in Bally (2005a,b); Mazza (2003); Streun (2003); Vogelsang (2004); Weigel (2003); Wirth (2004); Wunderlin (2004).

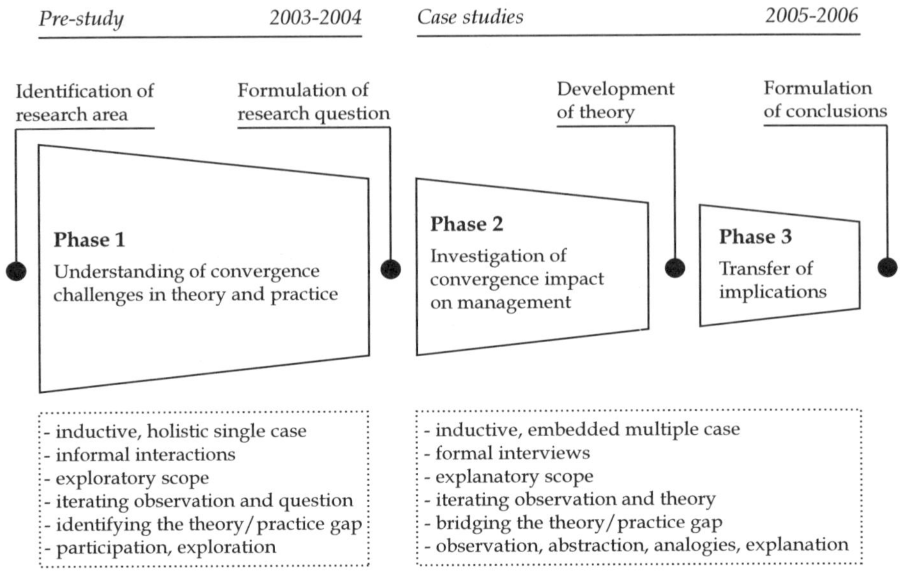

Fig. 1.5: Research process

study projects, additional secondary data was collected through surveying press releases and articles related to firms in converging environments, which served as a basis for forming the sample for subsequent study. Finally, data collection was accompanied by an extensive literature study, aiming at building a picture on the current research state in the intersection of convergence and innovation. On the one hand, this triangulation between several dimensions of field and desk research allowed an identification of the gap between shortcomings in theory and the call from practice (Ulrich, 2001). On the other hand, the exploratory approach based on the given data sources allowed a continuous iteration between observations and research questions.[16]

Although the research approach is designed as grounded and inductive, it is recommended to have an initial construct, i.e., an idea on what to look for in the empirical data, as well as a specification on *what* and *how* questions (Eisenhardt, 1989a). Based on the outcomes from the pre-study, the initially identified research area was specified through the formulation of an *a priori* construct, as well as research questions Q_1 to Q_3. Phase 2 of the study consisted of the embedded,

[16] Main results of the pre-study are reported in Hacklin, Raurich, and Marxt (2004a, 2005b).

multiple-case study within the two distinct samples. Using the inductive replication logic, the sample of 26 major ICT firms was studied in order to gain an in-depth understanding on implications of the convergence phenomenon on firms, as well as on approaches for dealing with the effects of the phenomenon in innovation management practices. The study was conducted through formal interviews, with informants in middle management, senior management and executive levels (table A.3, p. 213). Interviews were conducted by one investigator on a semi-structured basis, following a case study protocol (Yin, 1994). The protocol consisted of an initial background research of each firm based on mainly archival data, followed by an illumination of the case-specific convergence context during the interview, as well as an optional follow-up.

In phase 3 of the study, the induced theoretical framework was applied onto emerging trends of nanotechnology. In particular, a sample of 9 firms was studied, with the aim of transferring insights and externalizing the model into the formulation of recommendations. Apart from interrogating informants from the firm sample, stakeholder interviews with industry experts where conducted, aiming at gaining context know-how, thereby achieving more robust knowledge on the external validity of the findings. Hence, the study from previous phase 2 was not replicated, but the induced model was discussed on an iteratively inductive and deductive basis. Interviews within this phase were less structured than in the second phase, and were conducted with firms' founders or owners, senior management, executive management, as well as with experts mainly from academia (table A.4, p. 214). Also in contrast to the previous phase, interviews in phase 3 were conducted by two investigators.

In both phase 2 and 3, interviews with both samples were conducted during spring and summer term 2006 at Stanford University. Primary data sources included the formal interviews, as well as follow-ups by e-mail. Interviews were recorded and transcribed, which allowed a subsequent comparison on a cross-document basis, facilitating the recognition of patterns, and thereby the emergence and iteration of constructs among the replications. Collected interview data resulted in a total of 32 hours of recorded audio material, as well as 273 pages of transcribed annotations. Primary data collection was iterated until saturation at multiple levels of investigation could be observed, i.e., when marginal improvement became small (Eisenhardt, 1989a). Secondary sources consisted of archival data such as annual reports, press releases, websites, analyst publications, and newspaper articles, which were mainly collected in the year 2005. Hence, data could be triangulated between real-time observations and retrospective data, allowing the establishment of a chain of evidence, which in turn increases the construct validity (Yin, 1994). In either sample, both firm names, as well as names of informants have been anonymized.

1.5 Outline of the thesis

In order to map the nature of the previously discussed research process into a comprehensive written account, the remaining chapters of this report are structured as follows (cf. figure 1.6):

Chapter 2 aims at laying the theoretical fundamentals for the study. The current research state on convergence from a managerial perspective is discussed, and an initial determination of the phenomenon is provided based on previous work. A link between the convergence phenomenon, as well as technological change and innovation dynamics is further established.

Chapter 3 develops an evolutionary perspective on the phenomenon of convergence, building on the elaboration of previous research. Based on empirical observations, distinct phases of a convergence process are defined and elaborated. The resulting evolutionary frame on the convergence process serves as a foundation for elaborating managerial challenges related to the convergence phenomenon, which yield an induced theoretical framework.

Chapter 4 elaborates on capabilities needed for managing the convergence phenomenon within specific circumstances, thereby taking a step from challenge identification towards practical recommendation. The articulation and abstraction of such capabilities is grounded in the observations of the firms, and is continuously iterated with theory.

Chapter 5 translates the observed specific capabilities into managerial guidelines for navigating through the process of convergence. Guidelines are further articulated with respect to specific circumstances, and the results are discussed in the light of previous theory on technological cycles. Finally, the resulting model is transferred and applied onto a new and emerging area of convergence, which yields a set of specific recommendations.

Chapter 6 concludes the thesis, discusses the results and presents an outlook for further research.

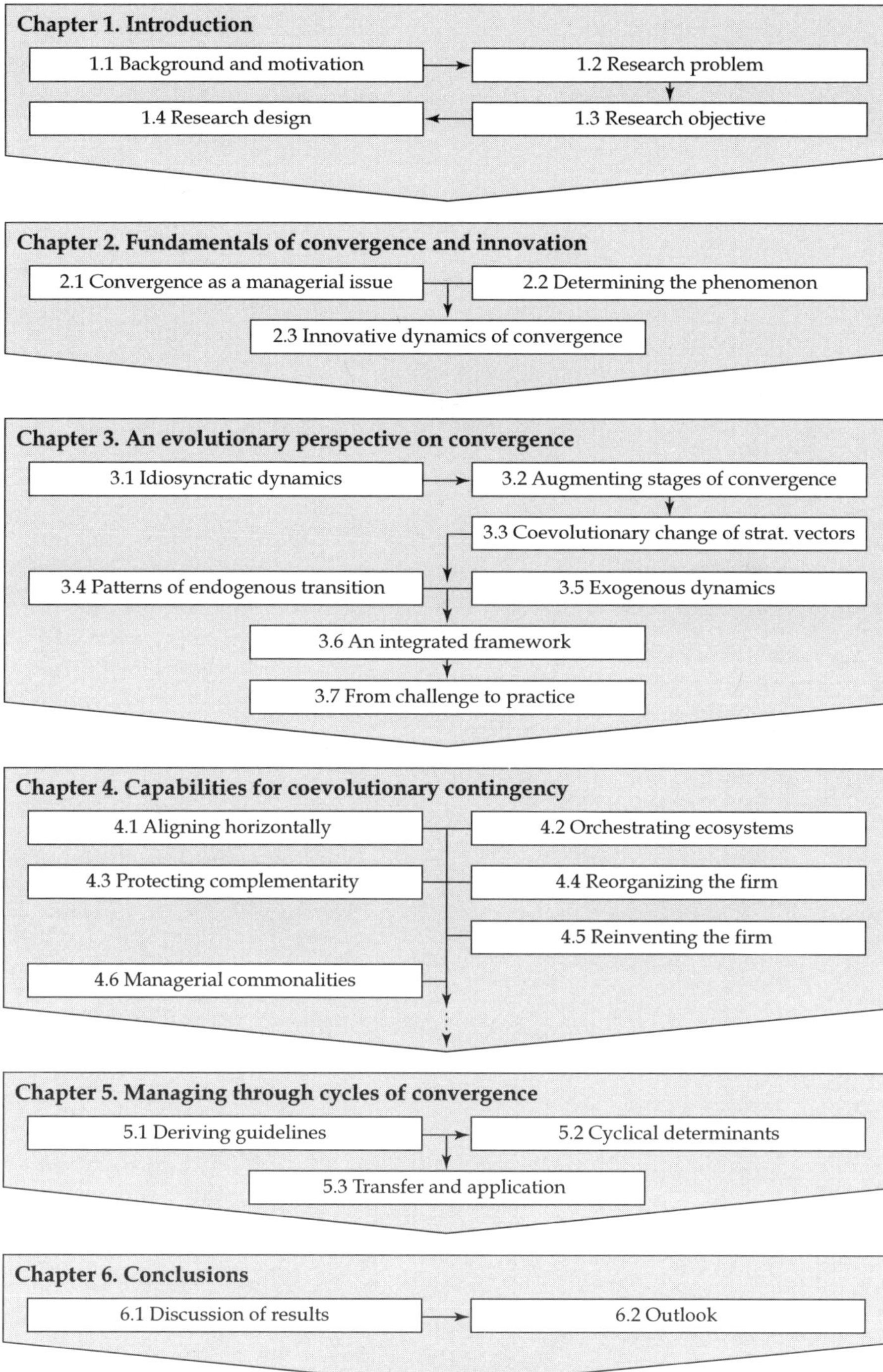

Fig. 1.6: Outline of the thesis

2

Fundamentals of convergence and innovation

Our current perception on technological convergence has been associated with the developments in the emergent ICT industry since the introduction of personal computers and the implied digitalization of media. In literature, this perception stems from the late 1970s,[1] when the emergent trend was already articulated through a *Science* article by Farber and Baran (1977) and identified as the convergence of computing and telecommunications systems.[2] Also, Nicholas Negroponte, the founder and chairman of MIT Media Lab, brought convergence onto the agenda in 1978 by illustrating a phenomenon through three overlapping circles moving together. These circles represented the three industries—computing, publishing and printing, as well as broadcasting and film. However, it seems that Negroponte missed the influence of the telecommunications industry at this stage (Brand, 1987). Also at that time, another similar visionary foresight on in-

[1] A broad overview on the usage of the convergence term in an early-ICT context is given by Lind (2004)

[2] Whereas Farber and Baran (1977) manage to identify the trends and to roughly predict implications in terms of future applications, they still conclude their discussion with a set of major problems to be solved prior to a common breakthrough. When considering the first three of the mentioned seven issues, one can today make the observation that, interestingly, all these mentioned issues turned out to become solved by the advent of the Internet:

"(1) The public availability of socially useful computer communications services is and has been held back by legal battles that are now under way between the potential suppliers. (2) No simple resolution of these issues in the near future seems likely in view of the past conceptual separation of computers and communications doctrines. (3) The current policy is to determine whether the nation shall or shall not have certain computer communications services, by the adversary process. In this process, often only the voices of the loudest adversary suppliers are heard." (Farber and Baran, 1977, p. 1169)

creasingly merging technologies within ICT was made in the concept of a 'wired society' (Martin, 1978), although the term convergence was not explicitly used. Similarly, an emerging overlap between telecom and datacom networks was addressed by a research project conducted at Harvard University in the mid-1970s, using the title 'Information Technologies and Public Policy', and introducing the terminology of 'compunications',[3] however, without further envisioning the impacts of a full convergence between these industry sectors (Lind, 2004; Longstaff, 2001; Oettinger, Berman, and Read, 1977).

In parallel, a call from practice for the emerging trend was generated by Nippon Electric Company (NEC). In 1977, the Japanese corporaton articulated a vision of digitalized integrated communication networks and distributed processing computers being converged by the year 1990. Based on this resulting 'C&C' (computer and communications) vision, the corporate strategy formulation was derived, resulting in a number of company mergers and aquisitions (NEC, 1984; Yoffie, 1996).

The intersection of technological convergence and innovation[4] can be regarded as the main underlying rationale for this research. This chapter aims at specifying the context, defining and explaining the phenomenon of technological convergence, as well as investigating state-of-the-art research.

2.1 Convergence as a managerial issue

When attempting to grasp an inclusive definition for the phenomenon of technological convergence from existing literature contributions, one may observe a certain vagueness in the current perception of the term (Katz, 1996). Particularly, the multitude of existing definitions in literature renders the comprehension of the term rather ambiguous (Theilen, 2004; Yoffie, 1996). Following up on this, in the work by Lind

[3] Referring to the word combination of *computer* and *communications*

[4] Throughout history, the term *innovation* has gradually evolved from a rather narrowly focused, technical term to a more open, commercially and socially related construct (cf. Barnett, 1953; Drucker, 1985; Fahrni, 2001a,b; Foster, 1986; Marquis, 1969; Marxt and Hacklin, 2005; Pavitt, 1980; Schmookler, 1966; Schumpeter, 1912). In particular, it is chosen to use the term for denoting the "process that begins with an invention, proceeds with the development of the invention, and results in the introduction of a new product, process or service to the marketplace" (Edwards and Gordon, 1984, p. 1). Already Schumpeter (1939) emphasized the importance of clearly differentiating between *invention* and *innovation*, which are "economically and sociologically two entirely different things" (p. 85).

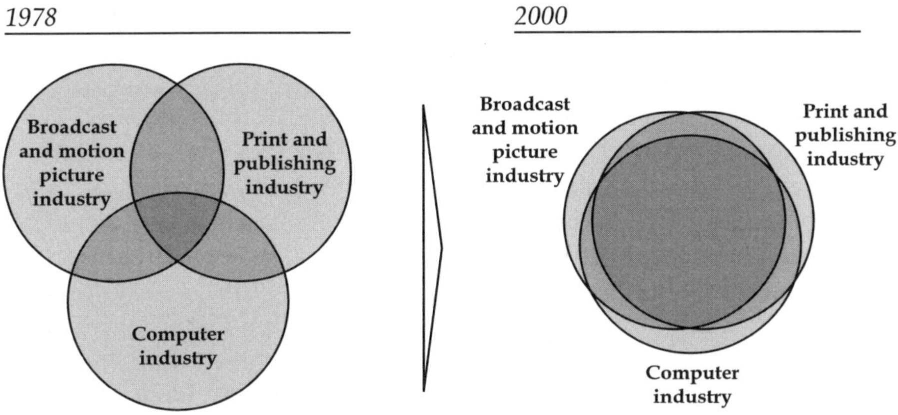

Source: Brand (1987)

Fig. 2.1: Early visions on convergence between industries

(2004), it is even argued that the concept of convergence has been commonly embraced without reflecting about definition or deeper meaning of the term and its implications. He furthermore infers that it is not only the business community who has ignored to do so—most academic contributions have neither tried to define convergence nor to relate it to a theoretical framework, but instead have taken the term as a given and just dealt with different implications of the phenomena.

2.1.1 Multitude of definitions

A commonly cited definition is provided by the Organization for Economic Co-operation and Development (OECD), explaining the overall phenomenon as the "blurring of technical and regulatory boundaries between sectors of the economy" (OECD, 1992, p. 13), and, implied from that as "the growing overlaps between the technologies, services and firms active in each sector" (OECD, 1992, p. 93). Similarly, another rather broad definition is given by Choi and Välikangas (2001, p. 426), who suggest a blurring of "boundaries between industries by converging value propositions, technologies and markets". Also, technological convergence can be seen as "the growing together of technologies, which fundamentally alters the boundaries of previously distinct industry or market sectors and merges them into a new competitive environment" (Bally, 2005a, p. 13). In an even wider perspective, convergence can be regarded as "a market/industry definition generated by technological change" (Lind, 2005, p. 1), and can be referred to as "processes that blur the boundaries and/or reduce the differences between activities, industries, nations and regions" (Bauer, Weijnen, Turk, and Herder, 2003, p. 2).[5]

Aiming at considering both internal and external pespectives, Thielmann (2000, p. 9) explains convergence as "a process of interaction between firm environment respectively competitive structure as well as corporate strategy, which leads to a structural connection between previously disparate markets".[6] Adding the consideration of distinct technology domains, Adner and Levinthal (2000, p. 64) refer to convergence as a process "in which the common domain is an application domain in which one of the two antecedent technologies is already applied".

Introducing a rather knowledge-related perspective, Pennings and Puranam (2001, p. 3) state that "convergence between previously disjointed markets can be viewed as the erosion of boundaries that define and isolate industry-specific knowledge". Based on that perspective, technological convergence is often defined as the process by which hitherto different industrial sectors come to share a common knowledge and technology base (Athreye and Keeble, 2000; Fai and von Tunzelmann, 2001; Gaines, 1998; Lind, 2004). Within this context, Guilhon (2001, p. 33) defines technological convergence as "the process by which the application of the same scientific concept allows the putting together of two or many fields of activity."

However, the most frequent use of the term *convergence* can be regarded as occurring in the context of ICT.[7] Being commonly referred to, the European Commission defines convergence within the specific context of ICT, namely as "the ability of different network platforms to carry essentially similar kinds of services, or the coming together of consumer devices such as the telephone, television and per-

---

[5] In a previous paper, Lind (2004, p. 1) introduces a working definition of convergence as "merging of hitherto separated markets, removing entry barriers across industry boundaries". In a later paper, however, he introduces the rather abstract definition, i.e., "a market/industry definition generated by technological change" (Lind, 2005, p. 1).

[6] Translated from German; original definition: "Prozess der Interaktion zwischen Unternehmensumwelt bzw. Wettbewerbsstruktur und Unternehmensstrategie, der zur strukturellen Verbindung bislang getrennter Märkte führt" (Thielmann, 2000, p. 9).

[7] Cf. Bane, Bradley, and Collis (1997); Barnes (2002); Bourreau and Gensollen (2004); Farber and Baran (1977); Harrison and Hearnden (1999); Messerschmitt (1996a,b); Yoffie (1997)

sonal computer" (European Commission, 1997, p. 1). More generally, convergence in ICT can be seen as "the unification of functions—the coming together of previously distinct products which employ digital technologies" (Yoffie, 1996, p. 33) and as "a process by which the telecommunications, broadcasting, information technologies and entertainment sectors [...] may be converging towards a unified market" (Borés *et al.*, 2003, p. 1). As an implication, this "dynamic approach or partial integration of different communication and information based market applications" (Wirtz, 1999, p. 15) results in "the blurring of borders between telecoms, computing and media" (Fransman, 2000, p. 39).

An overview of various definitions is given in table 2.1, containing a short summary in terms of industry and management scope for each source. Although the definitions mostly indicate the same, or even might be reformulating eachother, even recent literature does not completely leave out attempts for definition. However, the period of pioneering in defining technological convergence can be regarded as elapsed. In the following research, it is therefore chosen to focus rather on better understanding the phenomenon, than on introducing another generic definition.

Table 2.1: Overview of convergence definitions in literature

*Source**	*Definition*	*Industry scope**	*Management scope*
Adner and Levinthal (2000, p. 64)	"a process of convergence, in which the common domain is an application domain in which one of the two antecedent technologies is already applied"	general	innovation through recombination
Bally (2005a, p. 13)	"the growing together of technologies, which fundamentally alters the boundaries of previously distinct industry or market sectors and merges them into a new competitive environment"	general	industry and market boundaries; competition
Choi and Välikangas (2001, p. 426)	"[blurring] the boundaries between industries by converging value propositions, technologies and markets"	general	industry boundaries
Guilhon (2001, p. 33)	"the process by which the application of the same scientific concept allows the putting together of two or many fields of activity"	general	innovation through recombination
Lind (2005, p. 1)	"a market/industry definition generated by technological change "	general	technological change

Table 2.1: (continued)

*Source**	*Definition*	*Industry scope**	*Management scope*
OECD (1992, p. 93)	"the growing overlaps between the technologies, services and firms active in each sector"	general	competition
Pennings and Puranam (2001, p. 3)	"the erosion of boundaries that define and isolate industry-specific knowledge"	general	knowledge
Bauer *et al.* (2003, p. 2)	"processes that blur the boundaries and/or reduce the differences between activities, industries, nations and regions. [...] it is often used to denote the increasing similarity of regulatory regimes or a reduced variance in the performance of infrastructure services"	general	regulation
European Commission (1997, p. 1)	"the ability of different network platforms to carry essentially similar kinds of services, or the coming together of consumer devices such as the telephone, television and personal computer"	ICT	innovation through recombination
Wirtz (1999, p. 15)	"dynamic approach or partial integration of different communication and information based market applications"	ICT	innovation through recombination
Yoffie (1996, p. 33)	"the unification of functions—the coming together of previuosly distinct products which employ digital technologies"	ICT	innovation through recombination
Borés *et al.* (2003, p. 1)	"a process by which the telecommunications, broadcasting, information technologies and entertainment sectors (collectively known as ICT—information and communication technologies) may be converging towards a unified market"	ICT, media	market redefinition
Fransman (2000, p. 39)	"the blurring of borders between telecoms, computing and media"	ICT, media	industry boundaries
OECD (1992, p. 13)	"blurring of technical and regulatory boundaries between sectors of the economy"	ICT, media	industry and regulatory boundaries
Thielmann (2000, p. 9)	"process of interaction between firm environment respectively competitive structure as well as corporate strategy, which leads to a structural connection between previously disparate markets"	ICT, media	competiton and strategy

* Sorted by industry scope and source.

2.1.2 Development directions in research

The probably earliest use of the terminology "technological convergence" in literature can be traced back to the historical work of Rosenberg (1963, reprinted in 1976). In an analysis of the machine tool industry during the industrialization process in the USA, Rosenberg observes "the nature and the consequences of technological convergence" (p. 18) that exist throughout the machinery and metal-using sectors of an industrial economy, and identifies an emergence of common processes.

> "It is because these processes and problems became common to the production of a wide range of disparate commodities that industries which were apparently unrelated from the point of view of the nature and uses of the final product became very closely related (technologically convergent) on a technological basis—for example, firearms, sewing machines, and bicycles." (Rosenberg, 1963, p. 423)

In research, the issue of technological convergence has been mentioned in a variety of different contexts. For instance, Cameron, Proudman, and Redding (2005) analyze technological convergence from a productivity growth perspective. In particular, convergence is put into relation with technology transfer, which in turn is measured as statistically significant and important. Fagerberg and Verspagen (2002) examine the development of technological gaps between countries over time and study convergence or divergence of gross domestic product (GDP) per capita in the OECD area. The issue of convergence and interfaces between regional innovation systems are discussed by Heraud (2003) on a research policy level. Lemola (2002) concludes that instead of divergence there is an observable convergence of organizational forms and practices in national science and technology policies.

A significant contribution to understanding the phenomenon can be seen in the work of Lei (2000), who focuses on the strategic and organizational imperatives of technological convergence. In particular, it is examined how forces of technological convergence influence the development of industry structure. Using an industry-evolution perspective, Lei (2000) presents how technological innovations or changes emanating from buyers, suppliers, or firms in neighboring or related industries can pervade the economic relationships that span multiple industry boundaries.

> "As industry structures and boundaries become more permeable to the rising flow of innovations and new product concepts across different markets, these developementes often mean that technologies

commercialized in one industry could significantly influence, or even shape, the nature of product and process evolution in other industries." (Lei, 2000, p. 700)

Lei introduces a framework for examining the impact of technological convergence on the corporate strategies of a selected group of firms. In particular, he focuses on two distinct perspectives in investigating the impact: an industry-evolution perspective, that analyses effects of convergence on industry structure and boundaries; and a firm-based, competence-development perspective, that considers how technological convergence potentially redefines the firm's capability to learn and develop new sources of competitive advantage. Similarly, Collis *et al.* (1997) examine the general implications of digital convergence on industry structure.

The concept of "technology fusion" (Kodama, 1992; Teece, 1986) can be regarded as especially relevant to firms experiencing technological convergence (Lei, 2000). The concept is introduced as a higher-order approach to competence building in which firms search for ways to combine older, existing, and new technologies to create a blend of core competences. In particular, firms can attempt to merge multiple, incremental technical improvements from several previously separate fields of technology into hybrid products that revolutionize markets (Kodama, 1992, 1995). In the search for a delimitation from convergence, one might consider the observations of Adner and Levinthal (2000), who argue that in the case of technology fusion, the resulting technology is applied to a new domain. On the contrary, in the case of convergence it is the new domain itself—a new, resulting industry—that can be seen as consisting of a blend (cf. Day, Schoemaker, and Gunther, 2000; Gong and Srinagesh, 1996; Lei, 2000; Stieglitz, 2003). In the specific case of ICT, some scholars argue that convergence is not a complete fusion of existing industries, rather the emergence of a new market additional to the established industries (Baer, 2004; Gomes-Casseres and Leonard-Barton, 1997). On the other hand, Pennings and Puranam (2001) suggest technology bundling or fusion to be regarded as a specific archetype of convergence, and illustrate this assertion with examples from optoelectronics and bioengineering fields. In that context, examples for convergence developments, in retrospect looking like technology fusion, are introduced (Pennings and Puranam, 2001).

In the work by Baer (2004), the strategic phenomena of convergence and inter-firm cooperations are examined. In particular, interdependencies between converging industry sector developments as well as collaborative activities of firms are analyzed and confirmed. Based on this, firm-specific cooperation strategies and patterns are elaborated for the case of the ICT industry. In particular, Baer (2004) develops a conceptional model based on theoretic approaches for the analysis of cooperation portfolios in dynamic and converging industry developments, allowing the analysis of existing weaknesses within a cooperation portfolio as well as the goal-oriented complementing based on adequate collaborative arrangements.[8] Convergence developments between the single underlying industries, media, telecommunication and IT, and based on that, the analysis of the converging ICT industry, are illustrated and explained. A firm-specific empirical analysis of strategies and portfolios is conducted, allowing the structure of cooperation profiles together with its characteristics to be depicted, compared and evaluated. Besides comprehensive, generic insights, a correlation between convergence developments and cooperation strategies is observed. In this context, Baer (2004) argues that convergence developments cannot simply be extrapolated, but that they furthermore are depending on cyclical influences and fluctuations in the speed of development. The decline of cooperations and the slowdown of convergence developments since the end of the 1990s are mentioned to support this.

A generic strategic management perspective in the context of technological convergence has been addressed by various studies.[9] In particular, Kaluza *et al.* (1998) highlight the relationship between strategic management and converging industries by investigating implications on the critical success factors, i.e., cost, differentiation, flexibility and time. Referring to the five forces model for competition (Porter, 1980), a comparison between industry structure before and after convergence is made. As a conclusion, arguments on the strategic importance of both the five forces as well as critical success factors are made, although failing to present empirical evidence for this (Kaluza *et al.*,

[8] The evaluation model introduced by Baer (2004) is based on 11 parameters, each of which can be grouped into the categories of *convergence relevant* and *development specific* criteria. The convergence relevant criteria consist of: convergence intensity with regard to market focus of both co-operation and branch affinity; goal formulation and growth perspective of co-operation; changes of co-operation developments in the time; and coverage degree within the value chain. The development specific criteria are: degree of initiative; geographic focus; number of partners; power relations; functional focus; and formalization degree of co-operations.

[9] For example, Backholm and Hacklin (2002); Borés *et al.* (2003); Duysters and Hagedoorn (1998); Götte (2003); Hacklin and Marxt (2003); Kaluza *et al.* (1998); Kaluza, Blecker, and Bischof (1999); Messerschmitt (1996a); Pennings and Puranam (2001); Steinbock (2003); Wind, Mahajan, and Gunther (2002); Wirtz (1999); Yoffie (1997).

1998). Also, Wirtz (1999) introduces migration and integration strategies for cross-market dominance, which can be regarded as a further development of the resource-based view (Wernerfelt, 1984). Based on these strategies, companies can shift their core competencies towards new business areas emerging through convergence.

In terms of explicitly regarding and examining the convergence phenomenon as a process, there exists a relatively limited number of contributions, most of them being rather conceptual (cf. Mueller, 1999; Pennings and Puranam, 2001; Rockenhäuser, 1999; Stigler, 1951; Wirtz, 1999). Mueller (1999) suggests a long-term view of the convergence process, which is being supported by an analysis of prerequisites and impacts. Rather striving towards a holistic view, Stieglitz (2003) constructs a formal process or life cycle view for industry convergence, consisting of three stages:

> "In the first stage, two existing industries are unrelated from the viewpoint of both the supply and the demand side. A process of convergence is then triggered by an outside event, for example by the invention of a new technology. In the second stage, the industries converge, implying changes in industry boundaries, market structures and corporate strategies. Finally, in the third stage, the industries are related from a technological or product market perspective, and industry structures may stabilize, or new processes of convergence may evolve." (Stieglitz, 2003, p. 182)

Setting the issue of technological convergence into a context of innovation waves, Andergassen, Nardini, and Ricottilli (2003) investigate the evolutionary process of imitation and innovation as a process of searching in a given neighbourhood of firms. Translating this into the perspective of consequences, predictions of the coming of the digital revolutions were made during the mid-1990s, and attempts to assess implications for industry and society were made (cf. Baldwin, McVoy, and Steinfield, 1996; Yoffie, 1996)

The major contribution of Pennings and Puranam (2001) within this context can be seen in aligning the research towards a less abstract level of analysis, which consists of discussing firm responses to market convergence, in a manner that explicitly recognizes firm heterogeneity. Based on that perspective, it is investigated what strategic options are open to firms faced with market convergence, and a discussion of the firm-level attributes that determine the success in their strategic efforts is provided. Rather than examining the atypical instances of firms that successfully generate innovation, Pennings and

Puranam (2001) provide an analysis of a more *in vivo* situation of firms that attempt to deal with innovations that they have not originated.

In another study, an attempt is made for providing evidence to the assumption that technological diversification within firms, path dependency and technological convergence between industries are not necessarily opposing or conflicting forces (Svendsen and Fai, 2003). On the contrary, they can in fact work together in the same direction, perhaps in a sequential manner, which is being supported through the use of empirical data. Finally, relatively little quantitative research in terms of figures, ratios and other numerical indicators on the convergence phenomenon exist (apart from approaches such as introduced by Dowling *et al.*, 1998).

2.2 Determining the phenomenon

Greenstein and Khanna (1997) suggest two basic forms of convergence. On the one hand, convergence in substitutes occurs when the effect of the phenomenon causes redundancy and obsolescence between previously distinct domains, which leads to the effect of one *de facto* competing with the other. On the other hand, the case of complementary convergence occurs, when the coming-together of two previously distinct fields yields synergy effects, and results in more than the sum of its parts.[10] Building on this distinction, Stieglitz (2002) distinguishes between four types of market convergence, using the dimension of complementary vs. substituting aspects on the one hand, as well as product vs. technology orientation on the other. Also on an industry level, a convergence classification can be made based on scope of industry sector, where intrasectoral convergence developments occur within a branch, and intersectoral convergence developments between several branches of an industry (Baer, 2004). On the other hand, the phenomenon can be observed at different levels of value creation, since, "besides from convergence of technologies and infrastructures, convergent developments do even occur on the level of services and finally on the level of terminal equipment" (Baer, 2004, p. 303). Also contributing to a processual understanding, van Wegberg (1995) distinguishes among convergence on the supply and on the demand side. Convergence on the supply side occurs when industries increasingly use the same knowledge base, and demand-side convergence means

[10] In this context, convergence in substitutes, or competitive convergence, can be understood as "$1 + 1 = 1$", whereas complementary convergence has the effect of "$1 + 1 = 3$" (Dowling *et al.*, 1998, p. 32).

that market boundaries become fuzzier, both within industries and between them.

However, none of these attempts for providing classification to the phenomenon take into consideration the temporal dimension. In many cases, it remains hence unclear, whether the suggested different forms of convergence are mutually exclusive, to what extent they may coexist, and whether they occur in any sequential order.

2.2.1 Antecedents and implications

In order to develop an initial understanding of the convergence phenomenon on different levels, it is crucial to gain an overview of its various characteristics, as well as possible causalities between them. Instead of aligning managerial focus solely on dealing with symptoms (Lind, 2004), it is essential to investigate underlying reasons, such as "why do great companies fail" in the first place (Hamel and Prahalad, 1994, p. 128). Hence, in order to generate a more differentiated understanding of the phenomenon, there is a need to disentangle cause and effects of convergence.

Table 2.2: Drivers of convergence

	Endogenous drivers (firm-level)*
Andergassen *et al.* (2003)	– firms' search for innovative opportunities in other economic sectors
Lei (2000); Theilen (2004)	– new technologies that link up individual products into a larger system
Pennings and Puranam (2001)	– new technologies are discovered which map onto needs already being satisfied by existing technologies
Svendsen and Fai (2003)	– firm diversification and increasingly broader technological scopes
Yoffie (1996)	– start-up firms providing managerial creativity
	Exogenous drivers (industry-level)*
Andergassen *et al.* (2003)	– technical change, growth
Ibid.	– the greater the number of sectors, the more the same knowledge base is shared and the more specialized is the economy: hence, the greater the probability that information spills over
Antonelli (2001)	– emergence of pools of collective knowledge
Baer (2004)	– augmenting connectivity and interconnectedness provided through common standards
Ibid.	– several, increasingly multifunctional devices
Choi and Välikangas (2001)	– ubiquity of information

Table 2.2: (continued)

Ibid.; European Commission (1997); Lei (2000); Pennings and Puranam (2001)	– deregulation
Choi and Välikangas (2001); Pennings and Puranam (2001)	– growing similarity of demands across groups of consumers
Johansson (2004); Rosenberg (1963); Theilen (2004)	– underlying catalysator for information production and mutual transmittal
Lei (2000)	– growing opportunities for product bundling
Ibid.	– omnipresence of supplied product components
Ibid.; Theilen (2004)	– legal changes
Pennings and Puranam (2001)	– homogenization of customer segments due to changing demographics
Ibid.	– product bundling, customer demand for "one stop shopping", i.e., a desire to obtain in a single transaction a product that satisfies multiple needs
Ibid.; Petrina, Volk, and Kim (2004); Theilen (2004)	– globalization
Prahalad (1998)	– digitalization
Svendsen and Fai (2003)	– increasing complexity of consumer demands, dispersion of technological areas
Theilen (2004)	– market-implied changes of existing value chains
Ibid.	– societal change

* Alphabetically sorted by source.

When studying contributions from previous research for antecedents of convergence in technological environments, one can distinguish between drivers with either endogenous or exogenous source, i.e., either originating from within the firm, or from beyond it. On the firm level, a generic driver can be seen in deliberate search activities for innovative opportunities beyond current industry boundaries, where an existing technological base is diversified into broader scopes (Andergassen *et al.*, 2003; Lei, 2000; Svendsen and Fai, 2003; Theilen, 2004). Beyond the firm level, i.e., on an industry level, the general phenomenon can be identified as a confluence of a variety of different drivers. From a growth perspective, these comprise overall advances and in technology, spill-over between knowledge bases, and the increasing similarity of markets (Andergassen *et al.*, 2003; Baer, 2004; Choi and Välikangas, 2001; Pennings and Puranam, 2001; Theilen, 2004). On the other hand, socioeconomic developments and generic megatrends, such as globalization, digitalization, or societal change have be associated with the source of convergence (Petrina *et al.*, 2004; Prahalad, 1998; Theilen, 2004). Key aspects of identified drivers of convergence within existing literature are summarized in table 2.2.

Compared to the discussion of antecedents, previous research has rather extensively focused on the implications of convergence, which is also reflected in the summarized table 2.3. Similarly, one can observe effects of the phenomenon both internally, within the firm, as well as externally, i.e., within and between industries. In a firm-oriented perspective, entrants are suggested to primarily benefit from convergence through the emergence of new niche markets for breaking into established value chains (Chen and Hambrick, 1995; Lei, 2000; Yoffie, 1996). In contrast, incumbent firms may experience both chances and risks in converging environments, as current organizational capabilities can be rendered obsolete (Niemack and Weber, 2005; Pennings and Puranam, 2001). In particular, although convergence opens-up opportunities for creating new functionality and extending product features into new arenas, a deliberate managerial following of the convergence trajectory may cannibalize existing business (Theilen, 2004; Yoffie, 1996). Hence, new concepts and approaches for managing convergence on a strategic level are needed (Kaluza *et al.*, 1998), allowing not only methods for improved forecasting of emerging changes (Linton, 2002), but also enabling dramatic adjustments on various firm-internal processes to be made in time (Yoffie, 1996). On an industry level, the convergence seems to predominantly cause structural changes (Fransman, 2000), i.e., the blurring, erosion, and redefinition of previously established industry boundaries (Lei, 2000; Lind, 2004; Porter, 1985; Yoffie, 1996), thereby altering the competitive rules and systems (Dixit and Pindyck, 1994; Henten, 2004; Kaluza *et al.*, 1998; Lawless, Bergh, and Wilsted, 1989; Lei, 2000; Nelson and Winter, 1982; Yoffie, 1996). This leads to creative destruction based on the emergence of complementary offerings across industries (Lei, 2000; Markides and Williamson, 1996), which eventually can cause the appearance of an entirely new industry (Baer, 2004; Chakravarthy, 1993, 1994; Duysters and Hagedoorn, 1998; Kaluza *et al.*, 1998; OECD, 1992, 1996).

Finally, Pennings and Puranam (2001) note that convergence does not only have causes and effects, but is also often a self-reinforcing phenomenon. In line with this proposition, one can observe the occurrence of similar elementary aspects within antecedents and implications (cf. tables 2.2 and 2.3).

Table 2.3: Implications of convergence

Internal implications (firm-level)*	
Chen and Hambrick (1995); Lei (2000); Yoffie (1996)	– opportunities for existing and upstart firms to break into value chain based on new niche markets; challenging existing industry leaders for dominance
Kaluza *et al.* (1998)	– significantly alters the strategic management of four critical success factors (costs, quality, flexibility and time)
Lei (2000)	– need for internal R&D and strategic alliances across industries
Ibid.	– requirement of multiple dynamic routines that foster the creation of new sources of knowledge, insights, and capabilities
Linton (2002)	– need for forecasting disruptive and discontinous innovations
Niemack and Weber (2005)	– both chances and risks for established or incumbent firms
Pennings and Puranam (2001)	– degrade or render obsolete current organizational capabilities
Ibid.	– destruction of firm's coherent bundle of intangible assets; need to abandon a strict adherence to coherence, and to suspend the competency trap that comes with path dependence
Theilen (2004); Yoffie (1996)	– following convergence trajectory, albeit needed, may cannibalize existing business
Yoffie (1996)	– allow companies to create new functionality and extend product features into new arenas
Ibid.	– need for dramatic adjustments on firm-internal processes
Ibid.	– need for reaching large scale within core horizontal businesses, and find ways to leverage those economies into adjacent markets
External implications (industry-level)*	
Andergassen *et al.* (2003)	– development of new skills and competencies which creates a technological gap with other sectors
Antonelli (2001)	– emergence of technological clusters, i.e., a flow of complementary technological innovations which draw from a renewable pool of collective knowledge
Baer (2004); Chakravarthy (1993, 1994); Duysters and Hagedoorn (1998); Kaluza *et al.* (1998); OECD (1992, 1996)	– emergence of a new industry, e.g., ICT; on the basis of two several originating industries and the support of co-operation between firms of neighbouring industry branches
Bohlin (2000); Henten (2004)	– change of not only technologies, but services, markets, related actor configurations (industry alliances and mergers), policy and regulation

Table 2.3: (continued)

Dixit and Pindyck (1994); Henten (2004); Kaluza *et al.* (1998); Lawless *et al.* (1989); Lei (2000); Yoffie (1996)	– reducing the entry barriers and changing nature of rivalry; structural changes and growth in competition in the sense of an extrusion process; competitive conditions in which one industry's products or services are increasingly linked, absorbed, or blended with another industry's expanded range of offerings
Dougherty (1992); Dougherty and Hardy (1996); Lei (2000); Leonard-Barton (1992)	– alter the underlying nature of firms' core competencies, capabilities, skill sets, and underlying economic assumptions about the basis of competition in the newly configured industry
Fransman (2000)	– changes in industrial structure, technological evolution; fortunes of populations of firms in the industries affected by convergence, and by themselves
Guilhon (2001)	– facilitates the appearance of knowledge markets
Kaluza *et al.* (1998)	– increase in the bargaining power of buyers of the concerned industries
Ibid.	– increases the pressure from substitute products
Ibid.	– profitability of the converged industry will decline
Lei (2000); Lind (2004); Porter (1985); Yoffie (1996)	– blurring, steady erosion and re-definition of once distinct boundaries among industries as they begin to share more similar competitive, market-based, and technological characteristics; folding whole industries together
Lei (2000); Markides and Williamson (1996)	– creative destruction; giving rise to higher-performing substitute or complementary offerings that alter the nature of demand, competition, value creation, and firm behaviour among multiple industries
Rosenberg (1963)	– rapid generation of external economies of enourmous importance
Yoffie (1996)	– role of standards is further reinforced because of the threats of 'lock-in' and 'lock-out'; the installed base can generate excessive inertia and resistance to change

* Alphabetically sorted by source.

2.2.2 Multidisciplinary pervasiveness

Technological convergence has already been observed as a major element in the history of industrialization, where convergence occurred throughout the machinery and metal-using sectors of the industrial economy (Rosenberg, 1963), bringing together the steam and steel industries in the late 19th century (Lind, 2005). Today, the largest amount of examples of technological convergence is reported in the context of the ICT industry, such as e.g., the coming-together of the phone and the computer, which today

> "[...] both utilize digital technologies, but they historically served completely different markets with entirely different functions. The process of digital convergence implies that a computer begins to incorporate the functionality of a communicating device, and the telephone takes on the functionality of a computer." (Yoffie, 1996, p. 33)

In this context, the advent of the Internet can be regarded as the prime example of convergence, which initially started to blend computer and telecommunication areas, but later also began to encompass media and broadcasting industries (Henten, 2004). However, one may observe similar convergence phenomena in other branches, such as the emergence of machine tools combined with electronics, i.e., mechatronics, bioinformatics or the domain of leasing services (Baer, 2004; Lind, 2005). Prahalad (1998) argues that although the convergence of computing, communications, consumer electronics, and entertainment appears as a frequent example, the phenomenon is a lot more pervasive. Further examples for the coming-together of previously distinct disciplines are listed in table 2.4. Examples of underlying technological advances, resulting trends of technological convergence, as well as implications in the form of vertical disintegration are given by Pavitt (2004), and summarized in table 2.5.

When previously distinct techniques and disciplines come together and integrate each other, the resulting outcome will sooner or later be taken for granted, and one new industry may emerge. In the work of Anderson and Tushman (1990), the convergence of window and plate glass production techniques is observed as an example of two industries, which caused the formal unification into one single Standard Industrial Classification (SIC) code.

What among these examples however remains unclear, is the question of deliberateness. Whether some examples seem to stem from active managerial search for new combination across industry boundaries, other examples seem to originate from serendipitous confluence based on steady advances in science and technology. Also, the degree of change, and impact of the phenomenon on existing knowledge bases and products seems to vary among these examples.

2.3 Innovative dynamics of convergence

Introducing a fundamentally changing concept in evolutionary biology, Eldredge and Gould (1972) suggest that static lineages of an evo-

Table 2.4: Convergence between industries and disciplines

Disciplines	*Characteristics*
Personal care and pharmaceuticals	Pharmaceutical technologies and processes, including clinical trials, will increasingly be integrated into personal care products, e.g., shampoo and face creams.
Fashion industry and science	The nature of the personal care products industry will change its nature based on scientific advances in e.g., hair growth and anti-aging products.
Agricultural commodities and plant genetics	Agricultural commodity products such as soybeans, corn, potatoes, or cotton will need to contend with new developments in plant genetics, e.g., fertilizers, insecticides, or primary processing, bringing a new level of technological sophistication to a traditional business.
Material science, chemistry, electronics and software	Products such as digital cameras, printers and copiers are promising combinations of material science, chemistry, electronics, and software, where chemical and electronic technologies are being mixed.
Materials, electronics, software, mechanical engineering	Increasing amounts of combination of engineered materials, electronics, software and traditional mechanical engineering can be observed e.g., in the automotive industry sector.

Based on Prahalad (1998)

Table 2.5: Vertical disintegration in relation to convergence

Underlying technological advance	*Technological convergence*	*Vertical disintegration*
Metal cutting & forming	Production operations	Machine tool makers
Chemistry & metallurgy	Materials analysis & testing	Contract research
Chemical engineering	Process control	Instruments makers
		Plant Contractors
Computing	Design	CAD Makers
	Repeat operations	Robot Makers
New materials	Building prototypes	Rapid prototyping firms
ICT	Application software	Knowledge-intensive business services
	Production systems	Contract manufacture

Source: Pavitt (2004)

lutionary path might eventually become abruptly changed, through a "punctuated equilibrium". Building on that theory, Tushman and Romanelli (1985) establish analogies to innovation trajectories, by postulating that patterns of organizational stability and change can be characterized as punctuated equilibria. Similarly, Gersick (1991) concludes that punctuated equilibria imply an impossibility of radical change to be accomplished gradually and comfortably, furthermore leading to an eventual emergence of new organizational forms, that previously have not yet been imagined.

2.3.1 Innovation dichotomy and cyclical behavior

Based on the discrepancy between equilibrium and punctuation, the dynamics of innovation trajectories and required strategic change within the firm can be regarded as dichotomous (table 2.6). Innovation trajectories tend to evolve in cyclical appearances, continuously altering between stable and unstable states of the innovation system (Anderson and Tushman, 1990; Tushman and Anderson, 1986).

Table 2.6: The dichotomous dynamics of innovation

*Source**	*Stable state*	*Punctuated state*
Schumpeter (1912)	incremental	radical
Grossman (1970)	instrumental	ultimate
Normann (1971)	variation	reorientation
Maidique and Zirger (1984)	adoption	true
Yoon and Lilien (1985)	reformulated	original
Rothwell and Gardiner (1988)	reinnovation	innovation
Meyers and Tucker (1989)	routine	radical
Anderson and Tushman (1990)	continuous	discontinous
Bower and Christensen (1995)	sustaining	disruptive
Utterback (1994)	evolutionary	revolutionary
Schmidt and Calantone (1998); Song and Montoya-Weiss (1998)**	incremental	really new
Rice, O'Connor, Peters, and Morone (1998)	incremental	breakthrough
Johansson (2004)	directional	intersectional

* Sorted by year of publication; partly based on Garcia and Calantone (2002).

** Schmidt and Calantone (1998) as well as Song and Montoya-Weiss (1998) appeared in the same journal issue.

Further drawing on this insight, one can distinguish between two forms of punctuation within innovation cycles. A *technological discontinuity* can be regarded as the effect of innovations pushing forward the performance frontier along the parameter of interest by a significant factor, and by doing so striking existing products or processes at the very foundation of the existing technical order, as opposed to merely extending existing ones (Anderson and Tushman, 1990; Schumpeter, 1912). When technological discontinuities occur, prevailing states of incremental change are punctuated, and technological rivalry between alternative technological regimes is initiated. Instead of driving the process along current trajectories defined by a technological paradigm, an entirely new paradigm emerges (Dosi, 1982). Furthermore, the discontinuity inaugurates a new technology cycle, and initiates the era of ferment, where a variety of competing products, solutions or standards coexist, and the population of competing firms is growing (Tushman and Anderson, 1986). During this phase, rivalry occurs on the one hand between emerging new technologies, but on the other hand between old and new ones (Foster, 1986).

In contrast, another punctuation can be seen in the emergence of a *dominant design*, which terminates the era of ferment, and turns the cycle into the succeeding era of incremental change. In such situations, single architectures occur that establish dominance in a product class, marking a key event in the evolution of an industry through making the transition from a fluid to a specific state (Abernathy and Utterback, 1978; Utterback and Abernathy, 1975). Not only are sales figures likely to peak after the emergence of a dominant design, but particularly a shake-out among competing products and solutions will take place, resulting in a decreasing population of firms. Additional non-technological factors affect the development of a dominant design, which include collateral assets, industry regulation, strategic maneuvering at the firm level, as well as the interaction between producers and users. During this state, the technology base evolves incrementally until the next discontinuous advance overthrows it (Anderson and Tushman, 1990; Utterback, 1994).

Among various forms of discontinuous advances (cf. table 2.6 and Garcia and Calantone, 2002) one can further distinguish between two different forms affecting the existing knowledge base of an industry. In particular, a *competence-enhancing discontinuity* replaces an existing technology, by building on the know-how embodied within it. In contrast, a *competence-destroying discontinuity* occurs when the advance does not only replace a technology, but even renders obsolete the expertise required to master it (Tushman and Anderson, 1986). In other words, the punctuation impact of a competence-destroying advance seems more severe than in situations where an existing competence base is enhanced. This is reflected by the observation that the era of ferment following a competence-enhancing discontinuity is longer than the era of ferment following a competence-enhancing one. In turn, this can be explained by the appearance of many rival designs when a technology builds on a completely new knowledge base, where it will take longer for market forces to sort out these variants than if the knowledge base is enhanced (Anderson and Tushman, 1990).

Hence, the dynamics of technological change and resulting industrial evolution can be regarded as an iterative cycle. A stable state of retention, i.e., the era of incremental change, is punctuated through a technological discontinuity, inducing variation and growth within the population of technologies, products, solutions and firms. Hence, during the era of ferment, a stable growth of entrants can be observed, which however is terminated by the emergence of a dominant design, where selection, and thereby shake-out, starts to take place. Finally, the punctuation induced by the dominant design further transforms into a stabilized state, i.e., retention (Tushman and Rosenkopf, 1992, figure 2.2).

2.3.2 Disruptions and exogenous strategic change

Introduced by Christensen (1997), the term *disruptive technologies* refers to products, solutions and systems, which in their early phases aim to serve specific niche markets, but over time manage to outperform their competitors on mainstream markets (Bower and Christensen, 1995). Generally, established, incumbent firms rather tend to fail on disruptive technologies, whereas new-market entrants and upstarts enjoy the 'attacker's advantage' based on their relatively high risk-orientation and low path dependency,[11] although even the opposite has been shown (King and Tucci, 1999, 2002). In particular, "disruptive technologies then would be those technologies that render established technologies obsolete and therefore destroy the value of the investments that incumbents have made in those technologies" (Danneels, 2004, p. 248).

Regarding disruptive technologies, and—in a broader sense—disruptive innovation as one major driver for growth, prosperity and

[11] Cf. Afuah and Tucci (2003); Christensen (1997); Christensen and Rosenbloom (1995); Danneels (2004); Foster (1986); Henderson and Clark (1990)

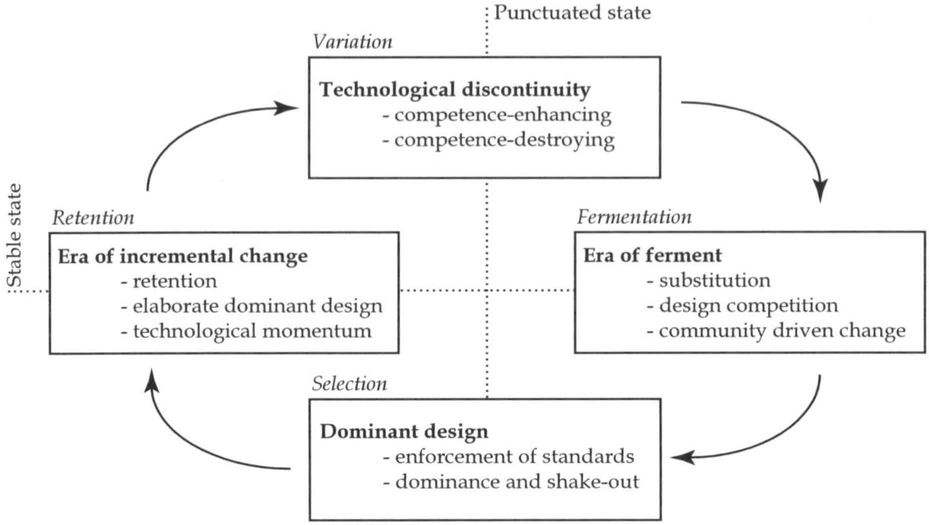

Adapted from Tushman and Rosenkopf (1992)

Fig. 2.2: Cyclical dynamics of technological change

wealth in today's hypercompetitive environment, recent scholars have been spurred to research on drivers, phenomena, and implications of such disruptions, aiming at understanding the dynamics and hence being able to catch the next wave in time. In particular, Paap and Katz (2004) see the major challenge for corporations in their imperative to manage both sets of concerns simultaneously. In order to achieve this, organizations need to learn how to manage the dynamics of innovation that underlie both disruptive and sustaining innovations. Whereas Paap and Katz (2004) further discuss the interaction between needs and technologies in this context, Adner (2002) identifies the demand conditions that enable disruptive dynamics, by constructing a model based on analyzing the intersection of customer preferences across different market segments. In earlier work, scholars have focused on the relationship between type of innovation and firms involved in successful commercialization.[12] In particular, they do not only suggest that disruptive technologies are more suitable for upstart firms in terms of market success, but they further suggest to in-

[12] Cf. Henderson and Clark (1990); Kassicieh, Walsh, Cummings, McWhorter, Romig, and Williams (2002a); Kassicieh, Kirchhoff, Walsh, and McWhorter (2002b); King and Tucci (2002); Kirchhoff, Kassicieh, and Walsh (2002); Tripsas (1997); Walsh, Kirchhoff, and Newbert (2002)

volve small firms for transferring new disruptive technologies from science labs to commercialization. Within a similar context, the evolutionary and cumulative nature of core capabilities and their interactions with technological discontinuities have been analyzed from a market-driven perspective (Walsh, Boylan, McDermott, and Paulson, 2005).

Based on such research, aimed at gaining insights into the phenomenon of disruptive change, a variety of tools and solution concepts have been suggested. For instance, Walsh (2004) introduces a general model for a commercial disruptive technology roadmapping process. More particularly, Kostoff, Boylan, and Simons (2004) present a roadmap for disruptive innovation, consisting of two parts, namely a text-mining component of literature-based discovery, as well as a practical workshop-oriented roadmapping process. Focusing on the integration into existing operative processes within the firm, Lichtenthaler (2004) introduces a concept for structuring the technology intelligence process in the context of radical technological change, and Trauffler (2005) proposes a multilevel process for integrating a goal-oriented management of disruptions into corporate strategy. As far as a sole identification or determination of disruptions is concerned, Vojak and Chambers (2004) introduce a heuristic methodology. Being based on observations of past patterns of change across several complex, technology-based, subsystem-level industries, it serves to identify potentially disruptive technologies. Based on a similar, but more theoretic approach, Linton (2002) introduces a mathematical formula set for forecasting diffusion of disruptive and discontinuous innovations. After all, new foresight techniques are needed, as in the peculiarity of disruptive situations, actions generally must be taken before careful plans are made (Christensen, 1997).

Whereas on the one hand it is regarded as questionable or even impossible to make *ex ante* predictions whether technologies are disruptive or not (Adner, 2002; Danneels, 2004), the work presented by Husig, Hipp, and Dowling (2005) makes an attempt to do so by developing a methodology for determination of disruptive technologies.

Danneels (2004) challenges and integrates current theory in the domain of disruptive technology, and initiates a discussion regarding related aspects for further research. In particular, he discusses various relationships of the disruptive technology work with research in a variety of related areas, of which many have not been made explicit before. In particular, Danneels (2004) states that one of the remaining questions is whether a technology is inherently disruptive or if *disruptiveness* is a function of the perspective of the companies subject to it. In the past, it has been shown that innovations which ultimately transform an industry often do not originate from the industry's leaders (cf. Cooper and Schendel, 1976; Danneels, 2004; Foster, 1986; Henderson and Clark, 1990; Utterback, 1994). Taking this argument one step further, on can argue that sources for disruption and technological change rather tend to originate from outside a respective system, which in fact already was observed by Rosenberg (1963) in the machine-tool industry. In other words, potential for disruption might occur when transferring innovation into new, previously unexploited market and competition environments.

Scaling such a phenomenon onto an industry level, such disruption potential could emerge from any case where boundaries between industries become blurred, as technology from one industry is adapted to serve the needs of customers in another industry. Particularly in converging environments, the emerging opportunities for applying their technologies to other markets drive companies to address different customers and offer different benefits than they had received before, paving the way for intersectional innovation (Johansson, 2004).

2.3.3 Impetus within the innovation dichotomy

The emerging obsolescence and substitution of technologies has already been identified as a common phase in the technology evolution process (Betz, 1993; Danneels, 2004), even though it was not originally and explicitly associated with inter-industry spill-over effects such as convergence. Applying technology life cycle-based views of such industry sectors onto this context, one can observe that the underlying technologies are often based on incremental, non-breakthrough improvements, which nowadays almost can be taken for granted (e.g., steadily increasing amount of computer memory capacity, data transfer rate, etc.). It is however the confluence of these incremental innovation processes that results in a larger-scale innovation with highly disruptive character (e.g., handheld wireless internet computing stations Hacklin *et al.*, 2004a, 2005b).

In literature, several indications for technological convergence representing a disruption in innovation trajectories have been given.[13] On the other hand, however, the underlying technologies often can

[13] Cf. Chandler (1997); Danneels (2004); Kapoor (2004); Kostoff *et al.* (2004); Lei (2000); Linton (2002); Markides and Williamson (1996); Prahalad (1998); Rosenberg (1963); Vanhaverbeke and Kirschbaum (2003); Yoffie (1996)

be regarded as well-known, established, and incremental. This problem has already indirectly been identified by Tidd, Bessant, and Pavitt (2005), who see one of the major challenges for current innovation management practices in managing both the discontinuous and the steady state, as breakthroughs can be triggered from clever use of existing technology in a new configuration for a newly emerging market. Based on this contrast, the question arises how this perceived disruptive change in terms of trajectories can be related to the underlying set of incremental technologies. Hence, in earlier stages of this research process, it is has been inferred that technological convergence can be seen as a special case of incremental innovation becoming disruptive based on confluence.[14]

Hence, such punctuations within stable innovation systems can cause new discontinuities, which are likely to imply change among firms involved (section 2.3.1). For the case of convergence, it is inferred that these implied patterns of change will need further translation and adaptation onto existing orientations and development paths of the entire company, i.e., strategic vectors (Burgelman, 2002).

As a consequence, the convergence phenomenon seems to cause a special type of impetus within the innovation dichotomy (table 2.6). In particular, this impetus causes a sudden change from a stable to a non-stable, i.e., punctuated state. The following chapter will aim at further analyzing the nature of this impetus as well as its consequences on firms and industries.

[14] These results have been reported in Hacklin *et al.* (2004a, 2005b) and are partly based on earlier work reported in Bally (2005a,b); Mazza (2003); Streun (2003); Vogelsang (2004); Weigel (2003); Wirth (2004); Wunderlin (2004). See also section 1.4.3.

3

An evolutionary perspective on convergence

Building on initial observations from the pre-study as well as literature review, this chapter will attempt at framing a construct for an evolutionary perspective of convergence phenomena, taking into account the effects on innovation dynamics and exogenous activities of a firm.[1] Based on the idea that technological progress is undetermined by factors internal to the technology, Tushman and Rosenkopf (1992) propose that it rather is the interaction of technical options with intra- and inter-organizational dynamics that determine the actual path of technological progress. This will serve as an underlying conceptual foundation, in developing an understanding of the convergence phenomenon in this chapter.

3.1 Idiosyncratic dynamics

In line with the proposition of Tushman and Rosenkopf (1992), convergence represents a multilevel phenomenon. On industry level, it is the development of underlying trajectories that drives the effects of change. These trajectories can hardly be controlled by any singular actor, and do emerge as a confluence of a variety of developments. On a firm level, convergence appears to be manifested in internal innovation activities, such as increasing development of bundled products

[1] In evolutionary biology, the phenomenon of convergent evolution describes the process of unrelated organisms independently acquiring similar characteristics while evolving in separate and sometimes varying environmental systems. For example, convergent evolution can be observed in the similar nature of flight behavior and wing characteristics between insects, birds and bats. In each of these cases, the wings are same in function and similar in structure, but each of them evolved independently (cf. Doolittle, 1994).

and solutions, which represent responses to the underlying industry trends and trajectory changes. Finally, somewhere in between, the level of inter-firm activities represent a major account for both firm-level and industry-level developments of convergence phenomena. This level consists of any form of exogenous firm activity and interaction, such as e.g., collaborations, partnerships, joint-ventures, mergers and acquisitions.

When examining the phenomenon of technological convergence from a managerial perspective, a variety of interesting, but complex effects of industry dynamics can be observed (Stieglitz, 2002). From a firm ecology perspective, the phenomenon on the one hand constitutes a major impetus for innovation and economic growth, while on the other hand leading to market disequilibrium and firm mortality (Pennings and Puranam, 2001). Similarly, rather idiosyncratic observations in terms of exogenous actions along the industry value chain can be made (section 1.1.2). Such idiosyncratic characteristics give rise to disentangling antecedents and implications on multiple levels, i.e., how actions of single firms affect the phenomenon, and vice versa, how the development of convergence affects managerial responses within firms (section 2.2.1). On the other hand, the temporal perspective, i.e., the process that happens between the antecedents and the implications remains of subject for managerial enquiry (section 1.2).

Hence, the convergence phenomenon renders peculiar characteristics of technological change and competitive dynamics. Since in a way, no firm seems to be directly responsible for convergence, on the other hand, *all* of them again seem to be, an understanding of when and how to act versus to react has to be gained. After all, research suggests that idiosyncratic situations can initiate change, thus firms may not be as inert as they sometimes are claimed to be (Ahuja and Katila, 2004).

3.1.1 A retrospective setting in ICT industry developments

For investigating such convergence-related dynamics, the recent transformations, involving everything between knowledge spill-overs and shifting industry boundaries, provide a relevant basis. Particularly the ICT industry provides an intriguing setting in this context, as it has developed rapidly and gone through several evolutionary stages in the very recent past. Not only was the industry life cycle disrupted by the boom of the 'dot com' entrepreneurial bubble as well as its succeeding bust. In parallel, a broad variety of inter-firm, and in particular, inter-industry activities such as consortia formation, joint-ventures as well as mergers and acquisitions has taken place. Strongly

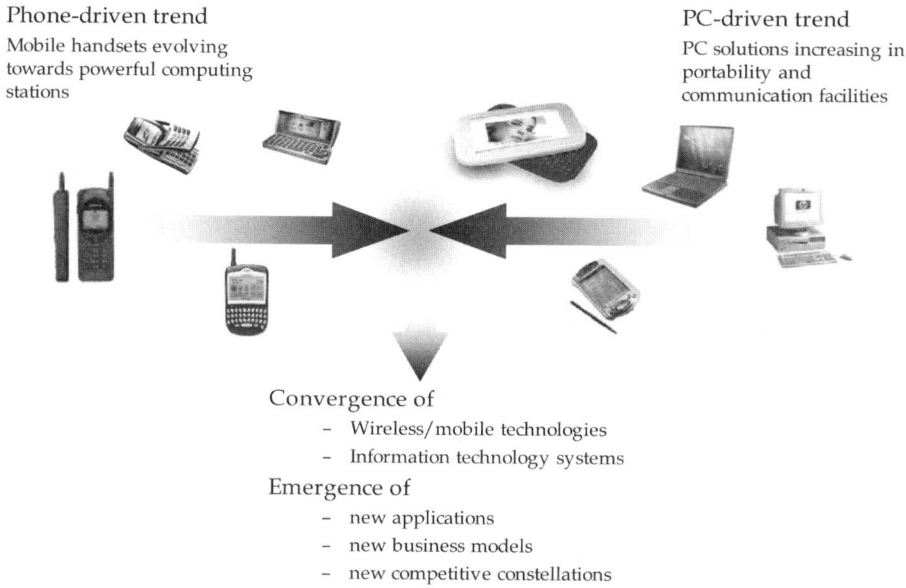

Source: www.nokia.fi, www.blackberry.net, www.hp.com, www.sony.com, www.umpc.com *(illustrations)*

Fig. 3.1: Convergence of previously distinct industries

interlinked with this development, the ICT industry of today features the stories of a variety of firms, which still ten or even less years ago acted in completely distinct fields, and which in turn today are positioned in mutually competitive constellations. Such significant alterations do not necessarily stem from the emergence of new market entrants, as even the trajectories of incumbent firms seem to lead to such competitive collisions.

Finally, the nomenclature of this industry *per se* displays an outcome of convergence, i.e., the coming together of *information* and *communication* technologies, which today implicitly is being taken for granted (figure 3.1). Based on this reasoning, the ICT industry represents an interesting basis for retrospective investigation.

Within the ICT industry case study, a sample of 26 firms has been studied. Whereas all the firms within the case set largely differ in terms of age, size and history, they share one common denominator: they are all acting in the intersection of distinct technologies or industries (Johansson, 2004), where the digitalization of information plays a crucial role. Based on the intersection of previously distinct technological domains, the firms in such a context see a potential to capitalize on the synergy provided by the confluence. As the informant 15 noted, achieving such a synergy, however, cannot be regarded as a trivial task:[2]

> "What you want to have with any convergence is the best of both worlds, you want more than the sum of the parts. And often, you end up with less than the sum of the parts." (informant 15, case β_1)

The emerging challenges of convergence, the managerial actions and strategic approaches represent the main level of analysis for the empirical material highlighted and discussed in sections 3.2 to 3.5.

3.1.2 Evolutionary elements of technological change

For analyzing the idiosyncratic dynamics that convergence implies in a broader retrospective, static approaches limit the modeling of such a complex process of change. Hence, the evolutionary theory of economic change will be applied as an underlying theoretical rationale for observing the phenomenon (Nelson and Winter, 1982; Nelson, Winter, and Schuette, 1976).[3]

Whereas neoclassical economics ignores the dynamic nature of change, and only focuses on the outcomes of the processes, the evolutionary theory of economic growth (Nelson and Winter, 1982; Nelson *et al.*, 1976) builds on the behavioral approach to individual firms (Cyert and March, 1963) as an interaction of firms with rather autonomously occurring external impulses. In particular, "the basic behavioral premise is that a firm at any time operates largely according to a set of decision rules that link a domain of environmental stimuli to a range of responses on the part of firms" (Nelson and Winter, 1982, p. 891). Hence, in the work of Nelson *et al.* (1976), the principal break with neoclassical tradition lies in the behavioral treatment of the question why a certain firm at any time is using the production technique it is using. Whereas a neoclassical answer would explain the choice of technique on the basis of profitability calculations comparing the elements of a large choice set, the behavioralist's answer is based on a different approach:

---

[2] This comment is in line with the elaboration on complementary convergence based on Dowling *et al.* (1998) in section 2.2.

[3] Scholars have previously built arguments based on relating the conceptions around technological convergence, spreading technological paradigms and broadening technological bases to an evolutionary perspective (cf. Cantwell and Fai, 1999; Dosi, 1988; Fai and von Tunzelmann, 2001; Lei, 2000; Sahal, 1985; Stieglitz, 2002, 2003; Teece *et al.*, 1994).

> "The production technique used by a firm at any time is regarded as a complex pattern of routinized behavior, of which the input-output coefficients are a quantifiable aspect. The firm is not seen as, at any time, "choosing" its technique from a large choice set, but rather as "having" its technique." (Nelson *et al.*, 1976, p. 94)

Evolutionary economics can be regarded as a relatively new economic methodology, which is modeled on biology and emphasizes on complex interdependencies, competition, growth, and resource constraints. Equivalent to genes in biology, firms are assumed to follow decision rules, or "routines", which together with impulses from the environment determine behavior. Such routines are heritable through "organizational memory" and may also change despite strong inertia, analogously to the biological phenomenon of mutation (Fagerberg, 2002; Lehner, 2000; Nelson and Winter, 1982).

Furthermore, Nelson *et al.* (1976) distinguish between two classes of dynamic mechanisms for describing the change process, that seem relevant in forming an evolutionary understanding on convergence:[4]

Definition 1. (Search) *The dynamic mechanism of evolutionary rule change that operates at the firm level is denoted as the process of search.*

At the firm level, rule change may occur as an implication of deliberate problem solving processes (such as research and development), possibly influenced by observed success and imitation of other firms. Also, it may "just happen" as a result of particular firm-specific capabilities improving through use, deteriorating through disuse, or adapting to changed input characteristics.

Definition 2. (Selection) *The dynamic mechanism of evolutionary rule change that operates at the market or economy level is denoted as the process of selection.*

At the market or economy level, the economic selection mechanism changes aggregate outcomes. Such a selection mechanism consists of the change in the weighting of different rules that comes about through the expansion of firms with profitable rules and the contraction of firms with unprofitable ones. Furthermore, the mechanism operates at any particular time, only on the set of rules actually employed at that time by functioning firms. However, this set is modified over time by the search activities of active firms as well as by the appearance of new entrant firms.

---

[4] See also table 3.1 on page 71.

As routines represent repositories and carriers of knowledge and skill, they represent an important object of study within business practices (Hodgson and Knudsen, 2004). Hence, evolutionary economics is essentially about changes in knowledge bases.

Finally, the evolutionary model can be regarded as focusing on the economic system acting between dynamics and stability (Dosi, Freeman, Nelson, Silverberg, and Soete, 1988; Klepper and Graddy, 1990; Nelson, 1995), as when forming cycles of technological change. As an implication, Nelson *et al.* (1976) argue that any impetus for radical shifts of foundations cannot result from analyzing long-run equilibria, further motivating the need for an evolutionary understanding.

3.2 Augmenting stages of convergent rule change

Building on earlier contributions from literature (section 2.1.1), the notion of convergence is regarded as an the intersection of existing established concepts, which in turn through their confluence render something novel. Given this rather broad abstractional scope of this notion, the phenomenon can be observed on a multitude of abstraction layers. In particular, initial reflections gained from the observation of current economic and industrial trends suggest such a multilevel impact of convergence to evolve gradually, i.e., reaching broader levels of scope as the trend evolves. As highlighted by Lei (2000), the convergence of technologies does not necessarily only introduce novel technological concepts, but might in many cases profoundly impact the evolution of industry structure.

In theory, scholars have contributed with a broad variety of convergence definitions and classifications in literature (table 2.1, section 2.2). In practice, the rather abstract phenomenon seems to occur in the form of several manifestations. Attempting at exemplifying these manifestations observed in the case set, and relating them to an evolutionary and sequential perspective, it is chosen to distinguish between the concepts as elaborated in the following sections.

3.2.1 Serendipitous spill-over between knowledge bases

Prior to taking an active step towards innovating at the intersection, an industry actor has to be able to identify a certain window of opportunity inside it. This is likely to be based on the awareness of a potential of combining the own, internal knowledge base with an external one, thereby creating something novel. Whereas such a step of combination can be made as a completely endogenous and conscious decision,[5] it is in many cases rather plausible to assume that an external influence motivates the industry actor to suddenly perceive the potential the way he does. Hence, one can regard the process of convergence as being ignited by the "the erosion of boundaries that define and isolate industry-specific knowledge" (Pennings and Puranam, 2001, p. 3), thereby decreasing the cognitive distance between previously distinct knowledge bases. Such an erosion of knowledge boundaries does not take place at the firm level, but rather through longer term developments of the industry. From a firm perspective, this can be regarded as a rather autonomous and serendipitous external effect. Hence, trajectories of knowledge bases come closer, and spill-over effects give rise to innovative activities.

Definition 3. (Knowledge convergence) *Knowledge convergence denotes the emergence of serendipitous coevolutionary spill-over between previously unassociated and distinct knowledge bases, giving rise to the erosion of established boundaries that isolate industry-specific knowledge.*

In the firm γ_1, an established player with a strong foothold in the semiconductor industry, the coming-together of knowledge bases on chipsets for central processing units (CPUs), as well as integrated circuits for communication technology, resulted in a completely new product architecture for personal computers. Whereas the company previously was committed on delivering the CPU as a core product, today's product leverages CPU and wireless communication capabilities on one chipset, with higher performance and less power consumption than separate elements used before. As the company saw this spill-over of previously distinct knowledge domains approaching, cross-organizational teams were formed and resources were allocated. When asked why such similar developments were not initiated earlier within the firm γ_1, e.g., through integrating CPUs with modem technologies into one single motherboard during late 1980s, the informants responded that it clearly was because of cost. In fact, not only have the costs of parts and components significantly come down since then, the company today experiences a cost advantage of integrating them. In other words, it is today even cheaper to integrate these technologies than to keep them separated. Previously, there were teams of engineers developing wireless communication technologies,

[5] Cf. Hacklin, Adamsson, Marxt, and Norell (2005a); Hacklin, Bergman, Nyström, Marxt, and Jantunen (2005c); Johansson (2004)

as opposed to others who designed the CPU chipset. As the underlying knowledge bases have become blurred, these people are today the same—a development, which the company was not directly able to influence, but rather respond to. And today, it would be much more costly and practically impossible to separate these knowledge bases.

Regarding its goal at helping companies innovate, the firm δ_4 designs products, services, environments, and—as the company names it—"digital experiences". Similarly as in the case of γ_1, the firm δ_4 responds to the converging knowledge bases through the formation of multidisciplinary teams in early stages of the product development process, for leveraging such competence synergies:

> "Although we might be working for, let's say, a consumer product industry, the team will have knowledge from health industry, from software industry, from service industry, so you can bring those things together and have more possibilities." (informant 6, case δ_4)

Similar trends of converging knowledge bases can be observed in the case of γ_5. By driving the coming-together of high-speed wireless and personal electronics into multi-use mobile devices, the company's technological solutions enable wireless phones, PDAs and laptops to support multimedia capabilities, such as high-quality video, digital cameras and MP3 players, accurate positioning and location services, as well as enterprise applications. In this context, as the boundary between the knowledge bases of communication and application capabilities became increasingly blurred, the firm decided to develop a common programming language, thereby bridging the worlds from an engineering perspective. Because the programming environment runs between the application and the chip operating system software, the application can use the device's functionality without the developer needing to develop code to the system interface or even having to understand wireless applications.

Essentially, the knowledge base of a firm can be disentangled into two domains (Granstrand, 1998; Teece *et al.*, 1994). Whereas one domain includes the knowledge about underlying technologies, building the technological capabilities, the other consists of rather specific knowledge about the market, its products, users, competitors, and applications, forming marketing capabilities.

Interestingly, whereas previously mentioned examples rather dealt with the serendipitous coming-together of previously distinct technological capabilities, the resulting need for market capabilities rather seems to diverge. In many cases, such as e.g., firms β_4 or α_5, it was perceived as a major challenge to find and recruit people who understand both underlying markets and businesses.[6] However, market capabilities can be regarded as complementary to the technological capabilities, as they enhance value generated by an innovation (Teece, 1986).

3.2.2 Technological opportunities for regime change

Strongly interlinked with the transition of knowledge bases, it has been argued that technological convergence can be regarded as a coincidental effect, where new technologies evolve through a confluence of random events, in combination with direct action of organizations shaping industry standards, as well as indirect actions implied by the competition of multiple organizations within the technological community (Tushman and Rosenkopf, 1992). Augmenting this model, the convergence of technologies can be observed as an autonomous, serendipitous underlying phenomenon, where previously distinct technologies increasingly start to share common technological properties (Rosenberg, 1963).

As knowledge bases eventually translate into technologies, this phenomenon, in turn, does not necessarily represent the result of any conscious managerial action, but can in many cases be regarded as a rather autonomous and complex process, which takes place beyond the firm level. It is as a consequence of the serendipitous coming-together of underlying trajectories, that new opportunities emerge, allowing firms to cross-fertilize throughout the technological intersection, and making technologies pervade new products. As the underlying trajectories converge, the involved technologies intersect in a way that a common technological knowledge base emerges, allowing opportunities for higher performance through diversification into new areas of within the underlying industries (Duysters and Hagedoorn, 2000). In particular, based on the converging knowledge base,

[6] In fact, a study by the recruitment firm Russell Reynolds Associates argues that convergence creates a major skills problem from an executive perspective. The survey results show that 65% of senior UK executives believe that convergence has created a skills gap, and furthermore indicate that there is a strong demand for employees demonstrating hybrid, cross sector expertise. In particular, executives perceive a need to acquire new skill sets for convergence, implying a need to look outside their traditional sector in order to get the right people. Currently, companies are not equipped to meet the staffing challenges and will need to develop a far greater awareness understanding of convergence and the industries that it spans (Williams, 2006).

basic design parameters which form the core of technological regimes (Georghiou, Metcalfe, Gibbons, Ray, and Evans, 1986) come to be increasingly similar as well, not only in terms of the material properties but also with respect to manufacturing processes involved (Duysters and Hagedoorn, 2000).

In other words, instead of considering technological convergence as the entire, time-spanning process, it is rather chosen to comprehend the phenomenon as the transition of knowledge convergence into a potential for technological innovation, which is arising through the fusion of previously distinct knowledge bases, and which in turn sparks higher level convergence phenomena (sections 3.2.3 and 3.2.4). Previously distinct technologies become hence suddenly related and are increasingly being addressed as a common set.

Definition 4. (Technological convergence) *Technological convergence denotes the transition of knowledge convergence into a potential for technological innovation, allowing inter-industry knowledge spill-overs to facilitate new technological combinations.*

An example for such a convergence of technologies and thereby emergence of intersected technologies can be seen in Rosenberg's (1963) observation of advances in metal cutting and metal forming techniques during the nineteenth century, which led to a potential for the coming-together of a number of manufacturing processes. In a more recent context, an example for convergence of technologies can be seen in the digitalization of information content previously bound to analogue storage and transmission technologies, which *per se* did not change the industry yet, but represented the prerequisite for extending and combining applications and business models beyond previously established boundaries (Lang, 2003).

All firms within the ICT case study base their innovations at least partly on a technological intersection, where the coming-together of technologies has created a tangible potential for the creation of new applications. For instance, in the cases of γ_1, or δ_7, β_5, the technological dimension can be seen in the coming-together of personal computing and mobile handset technologies. Computers steadily decrease in size, power-consumption and increase in communication capabilities, whereas cellphones have successively gained more computing power, memory, screen size, an so forth. Whereas the underlying knowledge base can be seen in *how* to process digital information, design chip sets or write application code, the technological dimension can be regarded as *what* parts and components eventually are being implemented and hence become available on the market.

While γ_1 rather today focuses its business on the convergence of the underlying technologies, δ_7 and β_5 build their businesses on further developed stages of the convergence evolution, which will be further exemplified below.

3.2.3 Application into synergy-leveraging business models

Once technological convergence is taking place, and a set of mutually intersected technologies has emerged, a window of opportunity for value creation emerges based on the idea of extending business models into wider areas of scope.[7] Once this opportunity appears, the innovator's actions become relevant for driving the convergence forward, as the trajectory of technological change now becomes dependent of the industry's ability to build upon the technological intersection. This step of integrating intersected technologies does not only result in the convergence into new applications, products or services (Robins, 2003; Stieglitz, 2004), but on a more generic level, leads to service or applicational convergence, as new, higher level forms of providing value for the customer and differentiation towards competitors emerge.[8] Hence, applicational convergence can be regarded as the emergence of an application vector, integrating the technology set.

Definition 5. (Applicational convergence) *Applicational convergence denotes the transition of technological convergence into opportunities for new value creation in such a way, that it with respect to the majority of metrics outperforms the sum of the original parts.*

In this application and service creation step of the convergence evolution, it is argued that the major managerial challenge consists in creating applications out of intersected technologies in such a way, which actually make them integrated. In doing so, future success is determined by a firm's ability to broaden its range of technological competencies, thus diverging away from the strict core competence notion (cf. Rao, Vemuri, and Galvin, 2004).

As an example, the emergence of camera phones, i.e., cellphones with an integrated digital still camera, such as manufactured by firm δ_2, makes an interesting case for the innovator's ability to create new service models based on convergence (cf. Tatsuno, 2006). Whereas the

---

[7] Cf. Ballon (2004); Kawashima (2002); Quinn (2005); Ralph and Graham (2004); Rolland (2003); Sigurdson and Ericsson (2003)

[8] Cf. Ciancetta, Colombo, Lavagnolo, and Grillo (1999); Edelmann, Koivuniemi, Hacklin, and Stevens (2006); Munoz and Rubio (2004); Steinbock (2005); Yang, Kim, Nam, and Moon (2004)

underlying trajectories, namely the photography technologies on the one hand, as well as the telephony technologies on the other hand, during a longer period of time already had begun to share an increasing common technology base, the step from taking this technological intersection into a technological integration which forms a new value proposition was needed. By introducing the camera phone, not only was a new product bundle created, but an even more far-reaching and disrupting service model for the way of taking and sharing pictures was introduced.

Similarly, in the cases of firms β_5 and β_2, the concept of mobile e-mail provisioning services represents an interesting application of the technological convergence between mobile communication systems and Internet technologies. Whereas the digitalization of wireless telephony in Europe was inaugurated through the global system for mobile communications (GSM) standard, the Internet started its pervasive diffusion during approximately a similar period of time. Although GSM in itself incorporates a mobile messaging technology, the short messaging service (SMS), there have been many entrepreneurial attempts to extend this messaging solution beyond being isolated within GSM networks, and bringing e-mail access into mobile handsets, thereby creating a new, broader user experience (cf. Grover and Saeed, 2003).

In such a context, the firm δ_7 has positioned its core business within this stage of convergence. As the technological convergence can be regarded as mostly explored and exploited by now, the company focuses on developing services that leverage the underlying advantages provided by the technological basis.

3.2.4 Shifting and redefining industry boundaries

When technologies are integrated and applications from previously distinct domains combined into novel applicational concepts, the effects may reach out beyond single projects or product generations. As emerging applications and service concepts evolve, they increasingly infringe the original value-creating territories of underlying sectors or industries. As a legacy of such a development, applicational convergence might lead to a collision of business models (Friedman and Waldman, 1992; Hackler and Jopling, 2003; OECD, 1992), as the development gradually removes the sectoral boundaries between the various involved industry segments (Duysters and Hagedoorn, 2000). Once initiated, such an effect of industry convergence can be pervasive, resulting in completely new competitive constellations, as previously established industry boundaries become blurred (Lee, 2003).

As an example, the coming-together of several, previously distinct technological bases within computing technologies, can be seen in and increasing blending of boundaries between virtually all major information technology markets, e.g., consumer electronics, broadcasting, instrumentation, military electronics, software, data processing, and telecommunications (Duysters and Hagedoorn, 2000). In particular, it has been recognized that industry convergence, based on the impetus of technological convergence, represents one of the major driving forces of economic developments in the international information technology industry, leading the industry to follow common trajectories (Duysters, 1996; Duysters and Hagedoorn, 2000; Forester, 1993; Georghiou *et al.*, 1986; Nyström, 2004). This has lead to a paradigm shift in the industry, where not only novel, improved applications were introduced within the technological intersection, but where basic assumptions have become significantly disrupted. For instance, whereas in the Internet pioneering years during the 1990s, circuit-switched telephone lines were used for connecting to the World Wide Web (WWW) using analog modems, today's paradigm can be regarded as vice versa, i.e., packet-switched technologies around the Internet are being used for carrying telephony services (Nyström and Hacklin, 2005).

In terms of competition, this paradigm shift changes the rules of the industry. For instance, while mobile handset manufactures and software manufactures in their origins—as well as in their business models—can be regarded as very unrelated, one can today observe direct competitive collision as the industry convergence is bringing along the battle for mobile handset software platforms. Such a collision of previously established business models resulting within converging industries, can in turn initiate consolidation according to the new rules of the competitive environment. Specifically, one might be able to observe alliances and M&A activity reaching beyond previously established industry boundaries.[9]

Definition 6. (Industrial convergence) *Industrial convergence denotes the transition of applicational convergence into the shift of industry boundaries in such a way, that firms from previously distinct industries through the emergence of common applications suddenly become competitors.*

[9] Cf. Bower (2001); Chan-Olmsted (1998); Collis *et al.* (1997); Dowling *et al.* (1998); Gong and Srinagesh (1996); Greenstein and Khanna (1997); Stieglitz (2003); Wirtz (2001); Yoffie (1997)

Examples of such trends can be seen in Internet companies taking over previously established media giants, as well as in incumbent telephony carriers distancing themselves away from their core telecom business with a succession of Internet-related acquisitions (Wirtz, 2001).

As far as the above mentioned camera phones are concerned, it is still unclear what impact this service concept will have on the industry convergence between digital still camera and cell phone manufacturers. In particular, one might see the current camera industry increasingly at risk, as their value creation model is being attacked by the growing technical capabilities of camera phones. In the long run, one might think of opportunities for consolidation and mergers between these intersecting industries, or shake-outs of entire previously established players might even be possible. Partly behind this trend, the firm δ_2 finds itself at such a commercial stage of the convergence process. On the one hand, the impact of new, convergence-orienting service models on the transformation or erosion of industry boundaries has become clearly crystallized. At the top management level, the firm is deliberately acting in a new, broader industry, with suddenly emerged competitors, from previously distinct industries.

Another interesting phenomenon of industrial convergence can be seen in the case of β_1. In particular, the entire foundation and commercial legitimation of the firm can be regarded as an effect of the maturing convergence process, as well as a response to a specific market vacuum. The implied structural changes of the carrier market derived a need for scalable, yet vertically-oriented and niche-focused business models (section 3.5.2).

3.2.5 A mechanism of cascading transitions

As a basis for further analysis, each case firm within the sample is associated to a specific stage of convergence, based on an analysis and comparison of currently observed main managerial challenges. An overview of all case firms associated to their respective stages of convergence is given in table A.5 (p. 215). In addition to the stage consideration, the table illustrates the underlying converging trajectories of each firm, as well as the firm's respective background and experience within those trajectories, denoted as the domain legacy.[10] Each positioning within the stage model is being motivated by a manifestation of the phenomenon, as observed in the data.

[10] In other words, the domain legacy represents a subset of the converging trajectories.

Since each of the previous models augment eachother, it is implicit due to the nature of this cascaded definition that knowledge convergence is inherent to technological convergence, which in turn is a precondition for applicational convergence. Finally, industrial convergence implies previous applicational convergence to have taken place. Hence, it is proposed as follows:

Proposition 1. (P_1) *The phenomenon of convergence can be comprehended as an evolutionary process of technological change, which is initiated through coevolutionary spill-over between knowledge bases of distinct industries, and can successively expand into more applied levels of convergence, which eventually can lead to the merging of entire industries.*

In particular, this successive expansion follows a mechanism of augmenting and cascading transitions that range through knowledge convergence, technological convergence, applicational convergence and industrial convergence.

Extending the formal notation as introduced by Stieglitz (2003) and Saviotti (1996) in a convergence context, one can conceptualize $\{\kappa_1, \kappa_2, \ldots, \kappa_n\}$ as a set of initially distinct knowledge bases. In such a setting, knowledge convergence implies the occurence of one or more mutual spill-overs within elements of the set, as well as the translation of these intersecting knowledge bases into a set of distinct technologies $\{\tau_{\kappa_1}, \tau_{\kappa_2}, \ldots, \tau_{\kappa_n}\}$, i.e.,

$$\begin{array}{cc} \{\kappa_1, \kappa_2, \ldots, \kappa_n\} \rightsquigarrow & \{\tau_{\kappa_1}, \tau_{\kappa_2}, \ldots, \tau_{\kappa_n}\}, \\ \text{with } \bigcap_{i\in\{1,\ldots,n\}} \kappa_i \equiv \emptyset & \text{with } \bigcap_{i\in\{1,\ldots,n\}} \kappa_i \not\equiv \emptyset \\ & \text{but } \bigcap_{i\in\{1,\ldots,n\}} \tau_{\kappa_i} \equiv \emptyset, \end{array} \tag{3.1}$$

where τ_{κ_i} represents the translation of knowledge base κ_i into corresponding technological capability τ (Georghiou *et al.*, 1986; Granstrand, 1998; Teece *et al.*, 1994). Based on such a state, potentials for exploiting technological overlaps based on knowledge spill-overs start to occur, and the process of technological convergence causes the technologies to intersect, and hence, to span an opportunity space. Accordingly, this spanned space can be modeled as a vector space $\langle \tau_{\kappa_1}, \tau_{\kappa_2}, \ldots, \tau_{\kappa_n} \rangle$, i.e.,

$$\begin{array}{cc} \{\tau_{\kappa_1}, \tau_{\kappa_2}, \ldots, \tau_{\kappa_n}\} \rightsquigarrow & \langle \tau_{\kappa_1}, \tau_{\kappa_2}, \ldots, \tau_{\kappa_n} \rangle \\ \text{with } \bigcap_{i\in\{1,\ldots,n\}} \kappa_i \not\equiv \emptyset, & \text{with suddenly} \\ \text{but } \bigcap_{i\in\{1,\ldots,n\}} \tau_{\kappa_i} \equiv \emptyset & \bigcap_{i\in\{1,\ldots,n\}} \tau_{\kappa_i} \not\equiv \emptyset. \end{array} \tag{3.2}$$

Followingly, through the exploitation and application of opportunities arising from technological overlaps, the stage of applicational convergence may lead to the emergence of new products, services, and business models, which are based on integrating the intersecting technologies, thereby leveraging the synergistic potential. Further building on the formal notation as used by Stieglitz (2003) and Saviotti (1996), such an integration can be represented as a vector $(\tau_{\kappa_1}, \tau_{\kappa_2}, \ldots, \tau_{\kappa_n})^T$, reflecting the notion of direction and impact of a multidimensional confluence. Hence,

$$\begin{array}{c}\langle \tau_{\kappa_1}, \tau_{\kappa_2}, \ldots, \tau_{\kappa_n} \rangle \\ \text{with } \bigcap_{i \in \{1,\ldots,n\}} \tau_{\kappa_i} \not\equiv \emptyset\end{array} \rightsquigarrow \begin{pmatrix} \tau_{\kappa_1} \\ \tau_{\kappa_2} \\ \vdots \\ \tau_{\kappa_n} \end{pmatrix}. \tag{3.3}$$

Ultimately, as nascent convergent business models increasingly start to reach out beyond previously established industry boundaries, a common base of technological capability emerges, which represents the stage of industrial convergence. This common base, denoted by $\tau_{\kappa_{1,2,\ldots,n}}$, consists of the union of single technological capabilities within the applicational vector $(\tau_{\kappa_1}, \tau_{\kappa_2}, \ldots, \tau_{\kappa_n})^T$, and encapsulates the convergent knowledge base $\kappa_{1,2,\ldots,n}$. Thus,

$$\begin{pmatrix} \tau_{\kappa_1} \\ \tau_{\kappa_2} \\ \vdots \\ \tau_{\kappa_n} \end{pmatrix} \rightsquigarrow \tau_{\kappa_{1,2,\ldots,n}}, \quad \begin{array}{r}\text{with } \kappa_{1,2,\ldots,n} \equiv \bigcap_{i \in \{1,\ldots,n\}} \kappa_i \\ \text{and } \tau_{\kappa_{1,2,\ldots,n}} \equiv \bigcup_{i \in \{1,\ldots,n\}} \tau_{\kappa_i}.\end{array} \tag{3.4}$$

Summarizing the formalized transition steps 3.1 to 3.4, figure 3.2 depicts the elementary mechanism of evolutionary convergence. An initial set of distinct knowledge bases undergoes knowledge convergence, resulting in the translation into technological capabilities, which *per se* remain mutually unassociated. It is the process of technological convergence, that turns the set of unassociated technologies into a span of opportunities, which allows the creation of products, services and solutions, based on integrated technologies, as rendered by applicational convergence. Finally, as this integration eventually becomes implicit, industrial convergence occurs, resulting in a converged technological base, which substitutes the set of original, previously distinct technologies.

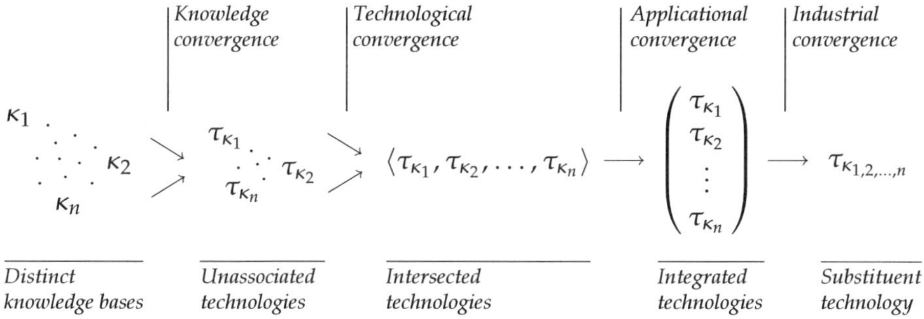

Fig. 3.2: Convergence as an evolutionary mechanism

3.3 Coevolutionary change of strategic vectors

Having its roots in organizational aspects of evolutionary theory (Aldrich, 1979; Baum, 1996; Baum and McKelvey, 1999), the coevolution of firms, i.e., the occurrence of interaction and interdependency between single evolutionary developments, can be identified both on intra-firm (McKelvey, 1999, 2002) and inter-firm levels (Burgelman, 2002; Porter, 1990, 1991). In particular, the emergence of intense coevolutionary behavior appears as an essential characteristic of modern competitive landscapes.[11] In the work of Longstaff (2001), it is argued that convergence may be understood as an outcome of firms evolving and adapting in relation to each-other.[12]

3.3.1 Classes of stage and maturity

Focusing on such an external perspective, the set of case firms can be segmented into groups of similar characteristics, as far as the basic setting in terms of strategic positioning with respect to the convergence challenge is concerned. Not only was it observed during data collection, that firms and interviewed informants start to make mutual reference to other firms within the case set, either in terms of competition or partnerships. When considering the case firms within their respective phases of convergent change (table A.5), and in combination with

[11] Cf. Bourgeois and Eisenhardt (1988); D'Aveni (1994); Eisenhardt (1989b); Eisenhardt and Tabrizi (1995); Galunic and Eisenhardt (1996); McKelvey (1999); Thomas (1996)

[12] "Convergence does not describe the process of coadaptation but, rather, is one of its possible outcomes." (Longstaff, 2001, p. 2)

their age or maturity, one can observe patterns of firms with similar characteristics (figure 3.3). Within this case set, the majority of younger firms studied happened to find themselves within rather intermediate phases of the convergence process, i.e., technological or applicational convergence. Similarly, the more mature and established participants seemed to be 'one step ahead' within the convergence process, i.e., covering technological, applicational as well as industrial stages of convergence.[13]

In juxtaposing both parametres against eachother, the stage dimension can be further simplified—by merging knowledge and technological convergence, as well applicational and industrial convergence, respectively—resulting in a distinction between early and late stages of convergence.

Definition 7. (Coevolutionary class) *A group of firms with similar observed parameters in terms of stage within the convergence process, as well as maturity of the respective firms, is denoted as a coevolutionary class.*

Hence, each coevolutionary class represents a specific configuration of a firm's maturity, as well as its strategic response to convergence challenges it faces at sequential times (cf. Rosenkopf and Tushman, 1994). The resulting classes of firms can consequently be identified in a 2×2 matrix, as well as labeled based on the overall strategic positioning of class members within the convergence phenomenon (figure 3.4).

Definition 8. (Pioneering disruptors) *The coevolutionary class consisting of firms with lower age than the median of the sample's firm age, and with a strategic focus on either knowledge or technological convergence, is denoted as pioneering disruptors.*

Definition 9. (Vertical attackers) *The coevolutionary class consisting of firms with lower age than the median of the sample's firm age, and with a strategic focus on either applicational or industrial convergence, is denoted as vertical attackers.*

[13] For simplicity reasons, the delimitation between entrant and established firms was drawn at the median of the year of foundation of all firms, which happened to be 1993. For the cases of δ_8 and δ_9, the positioning was manually changed based on the consolidation history of these firms. In particular δ_9 was privatized as late as 1997, which hence represents the year of foundation for the current corporation. Similarly, δ_8 was established in 1993, as a joint venture between larger groups. Based on the legacy inherited in both cases, both firms are in the following being considered as rather established than entrant.

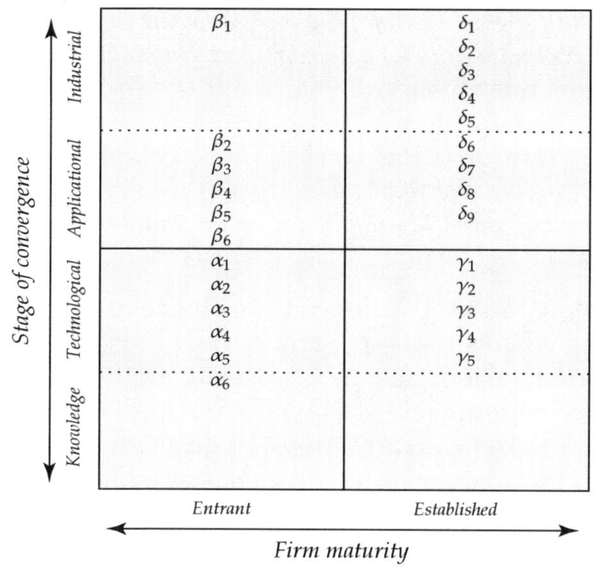

Fig. 3.3: Clustering ICT firms

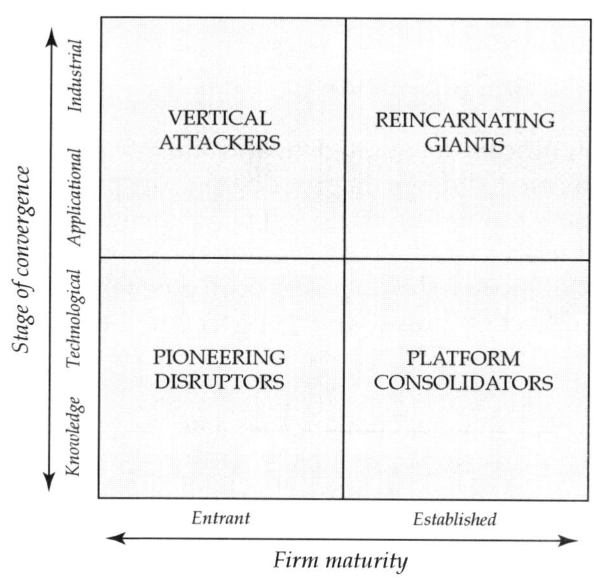

Fig. 3.4: Strategic positioning in coevolutionary classes

Definition 10. (Platform consolidators) *The coevolutionary class consisting of firms with higher age than the median of the sample's firm age, and with a strategic focus on either knowledge or technological convergence, is denoted as platform consolidators.*

Definition 11. (Reincarnating giants) *The coevolutionary class consisting of firms with higher age than the median of the sample's firm age, and with a strategic focus on either applicational or industrial convergence, is denoted as reincarnating giants.*

The discussion and motivation for the choice of nomenclature for these classes, as well as the appropriateness for each respective firm will not be further elaborated here. Instead, this identification and generalization of common classes among the studied firms will serve as a basis for the analytical part throughout the following sections.

By doing so, the generalization of such classes allows for more appropriate, abstract observation of coevolutionary aspects within different firms. Such coevolutionary groups, or pockets (McKelvey, 1999; Porter, 1990, 1991) can be contrasted with eachother, aiming at identifying industry-level dynamics in the context of convergence. Additionally, a direct and exclusive focus on firms for observing interactions and interdependencies, would probably soon come across the limitations of the small sample size in this study.

3.3.2 Inflection in firm orientation

In the face of significant technological and industrial changes, things happen to firms that did not happen before. In particular, the surrounding business environment no longer responds to the firm's actions the way it used to in the past, leading to situations where a firm can lose control of its own destiny. This phenomenon is being denoted as a "strategic inflection point" (Burgelman and Grove, 1996; Grove, 1997).

Convergence can lead to an overall point of inflection, i.e., some turning point of strategic change that can be associated to the serendipitous confluence of previously distinct knowledge bases. In particular, changes in strategic attitude or change on how to understand the role of oneself within the business environment seem to occur, often in the form of partnering, i.e., the way of dealing with the external world. Such strategic inflection points might either be related to competence-enhancing or competence-destroying technological change (Abernathy and Clark, 1985; Anderson and Tushman, 1990;

Table 3.1: Processes of search and selection

	Stage of convergence			
	Knowledge	*Technological*	*Applicational*	*Industrial*
Firm level process	Search induced by single actions			
Market or economy level	Selection induced by confluence of actions			
Focus of rule change	exploration pre-convergence		exploitation post-convergence	

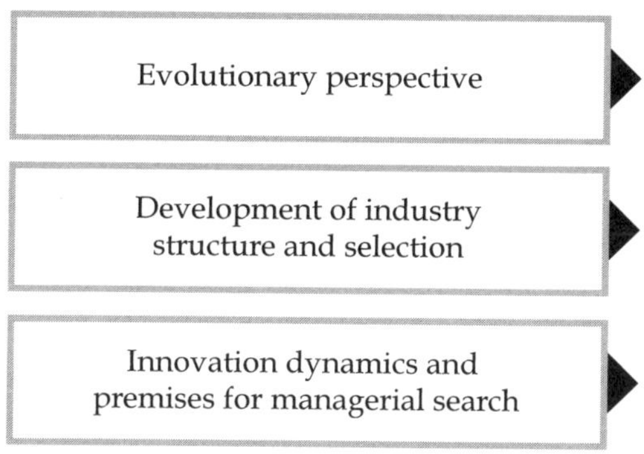

Fig. 3.5: Framework for model development

Tushman and Anderson, 1986). An initial overview of strategic inflection points among studied case companies is presented in table 3.2. These observations will we analyzed and discussed in further detail in the following sections.

The concept of strategic inflection points will serve as a level of analysis for sections covering *search* and *selection* (definitions 1 and 2), as further distinguished in table 3.1. In particular, inflectional patterns will be examined and compared in the observation of firms throughout the coevolutionary classes, aiming at constructing an integrated evolutionary framework (section 3.6). Hence, observations will be made on three levels (figure 3.5). Building on the previously introduced sequential view of convergence, the observations of different firms are replicated along both dimensions of the coevolutionary classes (section 3.2). Within the realm of these dimensions (i.e., firm maturity and stage of convergence), observations will on the one hand focus on the exogenous perspective of convergence, i.e., the way inflections stem from the industry and other factors external to the firm. In particular, this lens of analysis aims at gaining an understanding of the development of industry structure and selection mechanisms over time. On the other hand, as the observed phenomenon not only serendipitously coevolves between the levels of firms and industries, but also requires deliberate managerial responses, another lens will focus on understanding the observed innovation dynamics of convergence, together with the thereby resulting premises for managerial search.

In the following sections 3.4 and 3.5, observed evolutionary trends of convergence, will be distinguished and elaborated accordingly.

Table 3.2: Strategic inflection points

Case class	*Case firm**	*Implication on competence***	*Perceived severeness*	*Manifestation*
Pioneering disruptors	α_1	enhancing	low	development shift toward higher level applications
	α_2	enhancing	low	integration of new complementary business models
	α_3	enhancing	low	organizational changes for managing growth
	α_4	enhancing	low	internal change in technology base
	α_5	enhancing	high	refocus in customer target group
Platform consolidators	γ_1	enhancing	high	extension in technological scope
	γ_2	destroying	low	externalization of innovation
	γ_3	destroying	low	consolidation of customer base
	γ_4	enhancing	high	box selling to solution selling; verticalization
	γ_5	enhancing	high	development shift toward higher level applications
Reincarnating giants	δ_6	destroying	high	externalization of innovation
	δ_2	destroying	high	mindshift; diversification; externalization of innovation; reorganization
	δ_3	enhancing	low	partly opening-up vertically integrated model
	δ_4	enhancing	low	changing type of client projects
	δ_5	enhancing	low	exploration of new business areas
	δ_1	destroying	high	outperformance by convergent business models
	δ_7	enhancing	high	externalization of innovation
	δ_8	destroying	low	outperformance by convergent business models
	δ_9	destroying	high	mindshift; reorganization of innovation; reorganization
Vertical attackers	β_2	enhancing	high	increasing relevance of technology base
	β_3	enhancing	low	externalization of innovation
	β_4	destroying	high	refocus in customer target group
	β_6	enhancing	high	refocus in customer target group; externalization of innovation
	β_5	enhancing	high	increasing relevance of technology base

* Sorted by year of foundation (within respective class). Firms α_6 and β_1 are not listed, as the stealth mode in did not allow for such identifications.

** As defined by Anderson and Tushman (1990); Tushman and Anderson (1986), see also section 2.3.1.

3.4 Patterns of endogenous transition

According to evolutionary theory, rule change at firm level may occur based on conscious managerial action—such as problem solving or imitation—on the one hand, but also 'just happen' based on the evolution of capabilities (section 3.1.2 and table 3.1). In the observation of the convergence phenomenon, this section aims at identifying such implications on search, through highlighting its manifestations through several facets of internal change. In other words, the aspects to be spotted in this section, serve in the first place for understanding internal implications of the convergence phenomenon, as well as managerial challenges, which represent a basis for further identification of capabilities and development of managerial recommendations (chapters 4 and 5).

3.4.1 Self-induced competence destruction and lock-in

Despite entrepreneurial opportunities associated with technological convergence, many firms acting in converging environments face an idiosyncratic situation. The disruptive effect emerging through the coming-together of previously distinct knowledge bases can have a highly competence-destroying impact on firms involved in the transition. Since the emerging, convergence technology with its underlying knowledge base tends to substitute all previous antecedents, firms may experience remarkable difficulties in leveraging value based on their existing technological capability. In particular, this effect of 'out-innovation' can in many cases be observed as induced by the involved firms themselves—even though an actor within the technological intersection actively promotes the convergence of proprietary technologies with external ones, it is the actor himself who is disrupted at most through the convergence transition.

Particularly, firms within the class of *reincarnating giants* seem to be affected by this self-induced competence destruction effect (tables 3.2 and 3.3). Being themselves established, influential players within the industry, as well as on the other hand located within a mature, commercial phase of the convergence process, they attempt at gaining a share of the exploitation of business models arising through the intersection. Among these, firms of high path dependency with regard to technological capability seem especially vulnerable. Carrier firms, such as δ_1, δ_9 or δ_8, possess large infrastructure investments, and face major challenges in leveraging these competencies into value creation. One aspect of this 'carrier dilemma' (cf. Nyström and Hacklin, 2005; Saracco, 2005; Schneider, 2002) can be seen in the reaction to the emergence of open wireless communication standards, such as WLAN or WiMAX.[14] Whereas the advent of these technologies from the user perspective represent a major technological advance in the convergence of wireline and wireless communication systems, operators of communication services have been rather hesitant in their responses. For carriers, the migration toward this technological era would in many cases cannibalize investments made in previous technology. Particularly in Europe—and hence particularly in the cases of δ_1 and δ_9—significant investments in UMTS[15] have been additionally made, which in terms of application scenarios even could be regarded as a competing technology to WLAN or WiMAX. Whereas these carriers currently are struggling enough to achieve user adoption for UMTS, they regard it as rather questionable to commit resources in technologies than even further would cannibalize these investments.

Additionally, these open wireless communication systems additionally represent a strategic threat, as they due to their open and distributed nature tend to diminish the mechanism of control in terms of network operation or revenue creation, which requires further cautiousness when migrating proprietary networks into such systems. However, the firm δ_9 is operating a network of WLAN hotspots at public places, providing high speed wireless data access to its cellphone subscribers as a value added service, but does not regard the service as a reflexive threat to its existing offering.

Thus, in a convergent world, cores of the previously inimitable technological base might eventually deteriorate to rigid and substitutable legacies, allowing innovative efforts only in form of compromises. These hybrid models, in turn, might represent direct targets for the attack of innovative entrant firms.

But this effect can have a flip-side as well. Also, firms within the class of *vertical attackers* have a potential to end up in situations, where their own strategic actions increasingly out-innovate themselves. Acting in rather applicational and commercial stages of the convergence process as well, entrant firms such as β_2, β_5 or β_6 participate in the coming-together of applications, solutions and business models based

---

[14] Wireless Local Area Network (WLAN) is based on the standard IEEE 802.11; Worldwide Interoperability for Microwave Access (WiMAX) is based on the standard IEEE 802.16.

[15] Universal Mobile Telecommunications System (UMTS) is based on Wideband Code Division Multiple Access (W-CDMA) as the underlying standard, which is maintained by the 3rd Generation Partnership Project (3GPP).

on respective converging underlying technological bases. Whereas these firms in the short term actively strive towards facilitating this convergence, the business models can be regarded as intermediate and as potentially endangered, as the convergence development proceeds (sections 3.5.3, 4.3 and 4.6.3). For instance, developing middleware solutions for mobile e-mail such as β_2, can in the short term be regarded as a highly profitable solution in bridging the worlds of Internet and cellphones, but as respective established players start to develop similar capabilities, and the convergence becomes manifested, β_2 will have to refine that model. In fact, during the process of this study, the firm β_2 was already in the process of being acquired by δ_2.[16]

Hence not surprisingly, the firm β_1, entering the market in a mature convergence stage, does already at this early stage see a realistic exit strategy in the acquisition by an incumbent carrier or service provider in the—yet longterm—future (although informant 15 did not want to think about an exit too much at this moment).

Consequently, both pre-emptive action and forced reaction can, apparently regardless of firm size, lead to a rather peculiar effect of reflexive and self-induced competence destruction. Hence, it is proposed as follows:

Proposition 2. (P_2) *Along the process of convergence, a cautious transition from exploration to exploitation of knowledge is needed. Whereas exploration activities might lead to novel competencies on the one hand, these may at the same time induce inertial forces for exploitation.*

On the one hand, this specific inertia, manifested in a mixture of internal reluctance to change with external path dependency, may thus be interpreted as a special form of coevolutionary lock-in (cf. Burgelman, 2002). If actively working towards convergence eventually can lead to an inertia of coevolutionary lock-in, which in turn might result in out-performance through others, however, the question arises whether timing matters, and if, in particular, later market entry might result in more beneficial strategic situations, as one can wait and observe how the convergent technological base develops.[17]

---

[16] However, the case β_2 will in the remainder of this study be handled as a separate case.

[17] Such an effect would then be in contrast with existing literature, e.g., Suarez and Utterback (1995); Utterback (1994), who claim that entry prior to dominant design is associated with lower probability of failure.

Table 3.3: Self-induced competence challenges

	Stage of convergence	
	Knowledge *Technological*	*Applicational* *Industrial*
Competence rationale	exploration	exploitation
General competence challenge	novelty	inertia
ENTRANT FIRMS	PIONEERING DISRUPTORS	VERTICAL ATTACKERS
Specific competence challenge	development	lock-in
Example cases	α_6	$\beta_2, \beta_5, \beta_6$
ESTABLISHED FIRMS	PLATFORM CONSOLIDATORS	REDEFINING GIANTS
Specific competence challenge	acquisition	cannibalization
Example cases	γ_1	δ_1, δ_9

3.4.2 Opening-up of proprietary innovation mechanisms

As knowledge bases and technologies converge, location and scope of innovation processes tend to diverge. Regardless of firm maturity or convergence stage, all observed firms adjusted their internal innovation activities in response to the emerging intersectional scope. They experienced some kind of strategic change when it comes to dealing with the outside world of respective innovation processes. In most cases, the strategic policy for technological collaboration, seems affected.

In several entrant firms, represented by the classes of *pioneering disruptors* and *vertical attackers*, a trend of going away from fully manufacturing, deploying and marketing the innovations in-house was observed. Instead, these firms increasingly opened-up their innovation processes towards partnering with larger firms, aiming at combining their advantage of being small, agile and innovative, with the benefits of established firms' manufacturing, distribution and marketing capabilities.

> "[...] How do you move from being a 'one-product-one-technology'-company to being somewhat broader? [...] It would be naïve to think that we all by our self can come up with all and have monopoly." (informant 17, case α_5)

For example, for managing the company's growth and transition from niche positions into mainstream markets, α_5 is currently in the process of opening-up their in-house innovation process. For achieving higher diffusion of its fixed-mobile convergence products,

α_5 formed very close ties to γ_4 for co-developing product concepts and exchanging strategic views. Also, different middleware gateway providers and security providers are increasingly becoming involved in the innovation process, and even channel partners are involved in the assessment of market trends and opportunities. Based on this transition, α_5 is currently on a strong path from early adopters to broader bases of adoption (cf. Moore, 1991).

Whereas this form of opening-up appears rather strategic, the transition becomes more operative and substantial, as soon as the convergence turns applicational. In the case of β_3, a step on both fronts is perceived. Both β_3, as well as a set of partner firms have changed attitude when it comes to sharing innovation processes and associated value generation mechanisms. Hence, the firms innovation process is today opened-up towards a variety of different stakeholders in the system, such as hardware manufacturers of content distribution partners. For launching a mobile service product for accessing β_3's services from a cellphone, the firm partnered at early stages with the handset manufacturer δ_2, for allowing compatibility, reliability and a solid user experience. Hence, the service currently only works on certain models by δ_2, and interoperability has to be extended onto other handset platforms. But, as the informant 8 puts it,

> "You have to start somewhere. [...] We are strategically in the realm of leveraging a broader basis. Partnerships are the key here. [...] It's about building momentum, starting somewhere, learning from there." (informant 8, case β_3)

Again, β_1, entering the convergence process at a mature and commercial stage, builds its entire business model on collaborative and distributed innovation. The firm β_6, in turn, did not only move away from building all alone towards partnering with enterprises. This shift was accompanied by the change of business focus from business-to-consumer toward business-to-business.

Whereas the transition of opening-up proprietary innovation processes and systems in the cases of entrant firms seem to be associated with motivations for achieving product diffusion, exploiting established channels, or sustaining growth, it seems that even more serious attitude changes take place in the case of established firms. In such firms, the inherited legacy tends to lead to the fact that opportunities for true convergence remain somewhat constrained by the way that they are traditionally looking at it. The trick for overcoming such inertia and path dependency is in how to bridge.

Both the class of *platform consolidators*, such as γ_4, γ_2 or γ_1, as well as the class of *reincarnating giants*, such as δ_9, δ_6, δ_8 or δ_2 present manifestations of more or less large changes of mindset. In particular, these firms changed attitude on how to deal with smaller firms, and on how to benefit from their existence, instead of intrinsically regarding them as attackers and threats.[18]

> "Big companies have a problem, in that they force everything into their traditional metrics. The traditional metric for δ_6 is, how many pages are printed, how much ink is spilt. [...] They might miss in some sense the convergence, the true convergence. Because all they really see is more of an evolution, rather than a revolution. [...] versus if you're a smaller company, do you have resources to play with these [...] groups? But if you do, you are in the best position to think about how the convergence really plays out, because you are not framing it the way that traditional metrics were framed. " (informant 16, case δ_6)

In particular, those firms that find themselves at rather latter stages of the convergence process, i.e., the *reincarnating giants*, have undergone significant strategic reorientations at multiple levels. These have in many cases lead to heavy resource commitments onto organizing mechanisms for getting innovation from outside, by harnessing the entrepreneurial and innovative force of the variety of small firms out there.

For developing an online photo sharing service, which the firm δ_6 had based on its strong market presence in PCs, printers and digital still cameras in theory a highly suitable resource base for leveraging a sustainable business model. The product was developed in-house, shipped and evolved into its next version. However, despite huge engineering efforts, the company was not able to really crack the market. In parallel, a small start-up firm developed a similar product with sig-

[18] In particular, sometimes the emergence of new entrant firms within an existing realm can lead to erroneous conceptions by established firms. What at the first glace appears to be direct competition, can turn out to represent complementary growth from an incumbency perspective. In the case of δ_8, this resulted in the following insight:

"These others are not really competitors, rather complements. [Consider] small businesses who develop VoIP etc.—you have some specialization going on—and at customer premise, these small firms interpret applications in such a way, that new businesses SME begin to see the benefit of combining through a convergent platform. They have to and find a company like δ_8 [i.e., us] for handling and distributing traffic." (informant 9, case δ_8)

Table 3.4: Externalization of innovation horizon

	Stage of convergence	
	Knowledge *Technological*	*Applicational* *Industrial*
Control of innovation process	proprietary	open
Scope of innovation activity	internal	external
Dynamics of innovation activity	in-house; build-it-yourself	out-house; collaborative
ENTRANT FIRMS	PIONEERING DISRUPTORS	VERTICAL ATTACKERS
Innovation paradigm	one product, one technology	inside-out
Action rationale	compete	growth path
Example cases	α_5	$\beta_3, \beta_6, \beta_2, \beta_1$
ESTABLISHED FIRMS	PLATFORM CONSOLIDATORS	REDEFINING GIANTS
Innovation paradigm	one company, one innovator	outside-in
Action rationale	neglect	change of mindset
Example cases	$\gamma_4, \gamma_2, \gamma_1$	$\delta_6, \delta_9, \delta_2$

nificantly less resources, but seemed to succeed in the marketspace. Finally, the company decided:

> "I can't build it from within, no matter how many resources I throw at it. I'm going to buy it and acquire it. [...] The innovation needs to, in some sense, be seeded outside of the beast." (informant 16, case δ_6)

Whereas this represents one example of collaboration, which instantly lead to an acquisition, the firm has undergone huge internal changes of mindset during this period. Having a traditional internal R&D organization, with several large R&D groups and labs, the company started to work on externalizing innovation activities and overcoming the "not invented here, not created here"-type mentality.

At earlier stages of the convergence process, similar changes of mindset can be initiated, although the sense of urgency does not appear to be the same. In the cases of γ_2 and γ_1 for instance, decisions and resource commitments for opening-up innovation processes towards small firms have been made, although these changes seem to rather coexist with the remaining innovation culture at that stage.

A summary of these observations is given in table 3.4. Hence, regardless of firm size, innovation processes seem to have a tendency of opening-up under conditions of convergence. Hence, the following proposition is made:

Proposition 3. (P_3) *Along the process of convergence, an opening-up of proprietary innovation processes is needed.*

Whereas solid industry boundaries still may permit the innovation to be kept in a closed system, the shifting industry boundaries require innovation mechanisms to respond, thereby going from "cathedral to bazaar" (cf. Raymond, 1999, 2001). This tendency will be particularly highlighted in the context of external effects (section 3.5).

3.4.3 Inertial disciplinary structures and diversifying skill sets

Whereas section 3.4.2 describes the process of opening-up and partly deconstructing innovation processes and systems beyond boundaries of the firm, a similar parallel trend can be observed *within* the firm. Whereas previously distinct knowledge bases converge, and create a potential for exploiting and commercializing opportunities associated to the intersection, the horizon of knowledge bases and disciplines from a single firm's perspective has to open-up internally as well. On the one hand, actors in converging environments increasingly collaborate with the outside world in innovation, but on the other hand, they have to internalize more know-how from that external world. Whereas the evolutionary mechanism of convergence step-wise brings together technologies and eventually entire industries, hence reducing variance on a global perspective, firms in turn are increasingly challenged by multidisciplinarity.

> "It's not really convergence, it's collision. Whenever you have two industries that evolved independently and you're trying to merge them, you actually have a fundamental problem because of mindset. That's where it really starts. Mindset and language, people speak a very different language, people think in terms of different things." (informant 15, case β_1)

In observed case firms, the multidisciplinarity challenge was observed in two aspects:

- cross-disciplinary knowledge acquisition and integration,
- cross-disciplinary organization of innovation activities.

When industries increasingly come together, underlying disciplines have to be acquired into the firm. In most any case of ICT convergence, firms originating from the information technology side had to hire people bringing telecommunications industry expertise, and vice versa.

Providing a product that bridges the Internet protocol (IP) core to mobile access, α_5 is in the process of building expertise in both underlying areas. When the firm works with operators, GSM people have to be educated on IP technology, and vice versa. Hence, significant resources are invested in training within the company. But the firm regards this namely as one of the sources the firm brings. Nevertheless, α_5 tries to hire people with that skill set, which however is regarded as difficult.

> "So, to actually provision our 'box' to just make it work, you have to have IP expertise and GSM expertise. That combined skill set is actually very uncommon. [...] It makes things slow down a lot. [...] We are trying to bridge these two different technology domains together in order to offer the promise of convergence, but it's very painful, and the two areas do not merge without a lot of pushing. [...] It's a status quo, people are comfortable what they know already." (informant 17, case α_5)

In other words, this status quo, i.e., the functional fixedness of individuals from single disciplines, has to be overcome in order to facilitate innovation across these disciplines.[19] Similarly in firm β_4, the multidisciplinarity challenge implied by convergence was perceived to have an inertial consequence from a recruitment perspective. Particularly, several managers expressed a need for new education programmes and curricula for educating such hybrid skill sets.[20]

> "When you start doing that convergence, you simply don't find one engineer with all of these. He's not an RF [radio frequency] engineer, an electrical engineer, *and* a software engineer. I haven't seen one of those yet, when I do, I'm going to grab him!" (informant 21, case β_4)

Once required knowledge bases are in-house, either through external acquisition or internal training, organizational mechanisms for overcoming functional fixedness and internal 'not invented here'-syndromes have to be developed and applied. In this context, the issue of investing in corporate culture was mentioned as a crucial issue in several cases. In some cases, entire corporate structures might be reorganized in response to convergence, facilitating more, spontaneous, collaboration beyond previously static disciplines (section 4.4).

In the case of ICT, another interesting side-effect of cultural collision can be observed. When looking at the underlying technological streams, i.e., Internet-related technologies on the one hand,

---

[19] Functional fixedness is a cognitive bias that limits a person to using an object only in the way it is traditionally used (cf. Birch, 1945).

[20] See also footnote 6 on page 59.

Table 3.5: Trends of increasing multidisciplinarity

	Stage of convergence	
	Knowledge *Technological*	*Applicational* *Industrial*
Global knowledge bases (industry level)	multiple	converging
Local knowledge bases (firm level)	single	diverging
Disciplinary search processes	introverted	extroverted
Disciplinary task assignment metrics	static	dynamic
Base for expertise	within disciplines	between disciplines
Role of geographical clusters	protected	disentangled

as well as telecommunication, and in particular, wireless technologies on the other, a certain type of regional convergence seems to take place. Whereas the Silicon Valley area traditionally represents an epicenter for Internet-related technologies, the USA is generally being regarded as a laggard when it comes to wireless technologies (e.g., in terms of nationwide diffusion of handsets), where European markets appear more mature. On the other hand, European firms with their domain legacy in traditional telecommunications-oriented business areas, seem to open-up towards Silicon Valley peers for developing the knowledge bases related to Internet technologies. Hence, in many cases of the firm sample, the diversification could be seen in a transatlantic convergence, i.e., European firms focusing on geographical presence in the US, and vice versa.[21] Interestingly, the firm δ_2 appointed a board member from the US, heading Internet-related business activities—also based in the US—within the large telecommunications-oriented organization, thereby significantly strengthening not only its presence in the US, but thereby opening more doors for US firms in Europe. Hence, geographical clusters might intersect, and eventually partly disentangle.

A summary of multidisciplinarity trends is given in table 3.5. Based on these observed trends, the following proposition is made:

Proposition 4. (P_4) *As knowledge bases and higher level applications converge, involved firms' disciplinary horizons need to diverge.*

Whereas the multidisciplinarity challenge in many cases is perceived as a human resource management (HRM) issue, it may easily be extrapolated onto a strategic level. When industries with their underlying knowledge bases come together, it is crucial to organize mul-

[21] E.g., firms δ_2, γ_4, β_2, δ_9, δ_1, or γ_3.

tidisciplinarity issues beyond solely recruiting activities, otherwise the entire firm may suffer from limited absorptive capacity and limited bounded rationality (cf. Cohen and Levinthal, 1990; March, 1994; Simon, 1987).

> "For years, we were in the functional silos, and I think that we did everything we could to optimize the functions. And we recognize that the next level of productivity, and the next level of innovation, is coming from a cross-functional approach." (informant 19, case γ_4)

3.5 Exogenous dynamics and industrial inflections

Whereas the rule-changing process of search, i.e., the firm-level perspective, may contain visible elements of conscious managerial action or reaction, the inter-firm level, i.e., at market, economy or industry level, renders observations of much higher complexity, as they are embedded within the interaction of multiple stakeholders (section 3.1.2 and table 3.1). Hence, the economic selection mechanism can be characterized by serendipity, as a confluence of firm-level activity and related coevolutionary effects may result in industry phenomena, such as convergence. Nevertheless, within such a little amount of firms as in this sample, patterns of exogenous dynamics can be observed, which will be highlighted in this section.

3.5.1 Deconstruction of value generation mechanisms

At the stage of knowledge convergence, when underlying knowledge bases come together, the commercial phase is not entered yet, hence the business models are not intersected yet, and there coexist distinct and unassociated industry value chains (figure 3.6).

As soon as the technological stage of the convergence process is initiated, opportunities for new combination (mainly entrants) and diversification (mainly established) emerge. Hence, in total, more players providing same applications emerge, all of which, benefiting from the new economic advantage: the cost of integrating the knowledge bases goes below the cost of keeping them separate, since it is all encapsulated within the emerging converged—hence same—knowledge base, through dealing with the same knowledge base.[22] Hence, defining $C(x)$ as the cost of a resource x (either tangible or intangible), one can formalize this change as follows:

[22] Cf. example of γ_1 from section 3.2.1, with the integrated chipset, which would not have been possible in the old days.

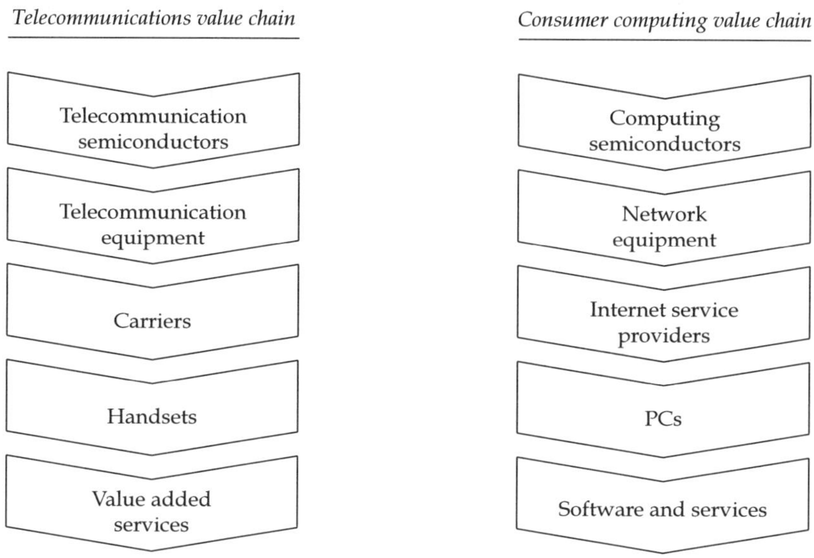

Fig. 3.6: Initial coexistence of distinct established value chains

$$\mathcal{C}(\kappa_{1,2,\ldots,n}) < \sum_{i=1}^{n} \mathcal{C}(\kappa_i), \quad \text{with } \kappa_{1,2,\ldots,n} \equiv \bigcap\nolimits_{i\in\{1,\ldots,n\}} \kappa_i \text{ (cf. eq. 3.4)} \tag{3.5}$$

In such a scenario, new market entrants aim at exploiting this potential mainly based on novel approaches. Firms within the class of *pioneering disruptors* have a tendency to do so by creating new combinations, thereby eventually introducing new technologies between established verticals structures. For instance, the firm α_5 serves as a horizontal bridge between network equipment and carriers, thereby creating opportunities not only for carriers' competitors, but also for the carriers themselves (cf. Ciancetta *et al.*, 1999; Yang *et al.*, 2004):

> "We initially actually through that our customer base are those who want to compete with mobile operators, and not themselves. [...] Today, we have gone into the other direction, as the value proposition is better for operators." (informant 17, case α_5)

Similarly, through introducing a broad variety of search engine-related applications, the firm α_3 is in the process of disrupting established vertically integrated value generation mechanisms through the advent of new search technologies, which in turn serve as a horizontal bridge between established structures.

> "Where this all becomes dicey for our partners is in the fringes of these industries where there *are* parts of the industry that have business models which [we], α_3 may make obsolete—if you extrapolate into the future. Almost always these are firms that aggregate the core players in an industry, because in fact that is what α_3 does. [...] Web search is a horizontal application because it covers basically anyone you want to search on. Same is true for α_3 book search, I consider that a horizontal search. You can imagine other searches across the broad body of human knowledge, maybe for a format or certain content characters." (informant 27, case α_3)

At this same stage of convergence, established firms have to tackle the phenomenon from the other side of the table. From the perspective of *platform consolidators*, one has to try to achieve spill-over through transferring and reapplying technologies between verticals. Being in the process of organizing such spill-overs, firm γ_1 is starting to leverage their existing bases of computing semiconductor technology for new telecommunication platforms. Similarly, in the case of γ_5, the diversification is originating from the opposite side, where knowledge convergence allows the firm's domain legacy in telecommunications semiconductors to be reapplied into mobile computing environments.

An even more evident dyad of diversification can be observed in the mutual interactions and developments of firms γ_4 and δ_2. Whereas γ_4 currently is undergoing a strategy vector of increasingly moving into the telecommunications domain, the firm δ_2, in turn, is currently moving into the opposite direction, i.e., network equipment.

> "When I was at γ_4, they were making more the moves into solutions based products, they were going through the same pains of 'how do we come a more software-centric company', and they went through or are going through similar pains as we are going through." (informant 4, case δ_2)

As the convergence process proceeds, and applicational stages of the process begin, consequences for previously established vertical structures occur. In particular, industry value chains become deconstructed, as new intermediate gaps for entrepreneurial actions emerge. Vertical disintegration occurs and involved players do not only step into an intersection of knowledge bases and technologies anymore.

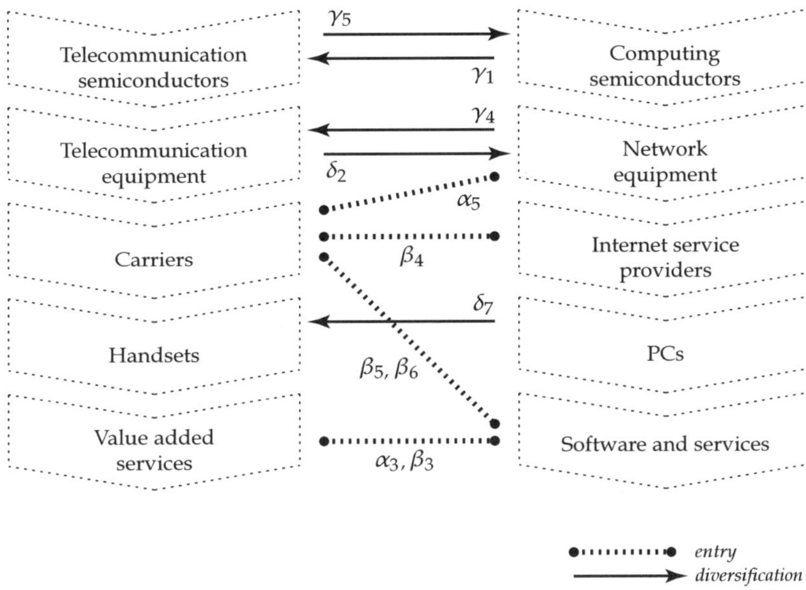

Fig. 3.7: Modes of spill-over between distinct business models

As soon as an overlap and collision of established business models emerges, the firms are thereby located in the intersection of entire value chains. Hence, established value chains become deconstructed, and vertically integrated systems disrupted. In other words, the industry structure becomes modularized, allowing new forms of recombination of structural elements into new business models (cf. Baldwin and Clark, 1997). Along with the proceeding into industrial convergence stages, players in established vertically integrated structures have to share the margins with horizontal attackers, such as mediating technologies or players who simply approach certain levels of the value chain from different sources. This happens based on either combination or diversification (figure 3.7), and results in a deconstructed, unorganized and entropic state of value generation mechanisms. In a way, horizontal and vertical becomes blurred, as some interdependencies might be vertical, whereas some spill-over occur horizontally, as e.g., in the previously mentioned example of firms γ_4 and δ_2, who find themselves in a dual role of relationships: being mutual customers and competitors at the same time.

However, as the vertically disintegrated state can be regarded as a non-stable state, a reconfiguration of value chains is being initiated (cf. Cacciatori and Jacobides, 2005; Jacobides, 2005; Li and Whalley, 2002; Wirtz, 2001; Yang *et al.*, 2004). In particular, one can see an emergence of value networks spanning over several previously incumbent value chains as a consequence of attempts of vertical reintegration and industrial consolidation. Again, the firm δ_2, who within a different division of the company currently is in the process of integrating digital camera technology into their mobile devices, is undergoing a diversification into the camera industry. From the perspective of established camera industry players, this causes a major external disruption, and it is still to be awaited to what extent the camera industry will be able to exploit the convergence of these knowledge bases as well, i.e., by integrating phones into their cameras. As a consequence, major challenges for firms consist of the internal fragmentation of business models based on diversification. In particular, incumbent firms tend to experience difficulties in covering several convergence niche areas at once.

Hence, this stage of convergence brings along new forms of interfirm interaction, i.e., consortia formation, clustering, and collaboration within standardization bodies. All in all, the impetus of reintegration of industry structures leads to network mechanisms (figure 3.8).

Looking into the future, firm α_6 bases current strategic activities on similar future effects of the convergence process, which are still to come.

> "A lot of our value proposition is based on the idea that, as the technology matures, on a lot of horizontalization occurs. [...] It is very hard to build a platform in a vacuum. We are architecting our products so that they can become a platform. [...] Our customers might not realize they need a platform yet. [...] We're overly invested in working with standardization bodies for driving interface specifications, that make horizontalization possible. Significant amount of manpower is working on pushing clean architectural layers and standards." (informant 13, case α_6)

Referring to the firm's competitors, informant 13 further notes:

> "The problem they run into, they are in version 1.0 of the convergence market. We're aiming our product development into version 2 or 3, where you have horizontal layers, do have some specialization across the links in the value chain. Version 2 or 3 has a very different business model in terms of what the operators want to buy, than version 1 model so, even though we aren't directly competing, theres a business model that were betting on, that might replace their business model, which puts us into favour as an independent horizontal vendor." (informant 13, case α_6)

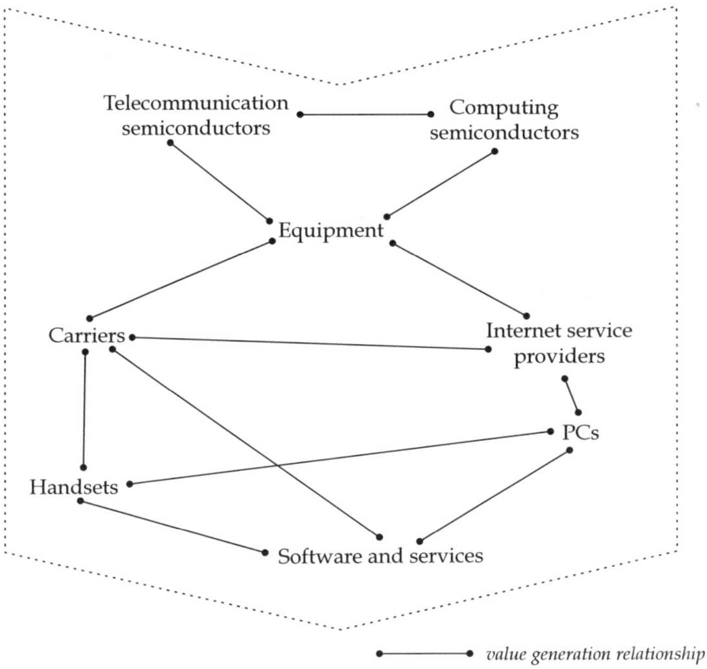

Fig. 3.8: Disintegration of value generation into network

In such stages of the convergence process, entrant firms, i.e., *vertical attackers*, tend to infringe margins of their established competitors, and attempt to substitute business models through new horizontal approaches. Both working in similar applications around search technologies, the firm β_3 distinguishes itself from firm α_3 as it finds itself somewhat more in the commercial phase of convergence. Hence, whereas firm α_3 still is dealing more with the technological level (see above), firm β_3 is currently specializing into a variety of specific business models. Similarly, the firm β_4 is bringing business models of carriers and Internet service providers together, thereby attacking carrier margins. Also, firms β_5 and β_6 develop solutions for allowing computer software and services to be integrated into carrier offerings, thereby outperforming some carriers within their own domain.

For established firms in this stage, i.e., the *reincarnating giants*, the challenge consists in leveraging their incumbency to regain the margins that were lost through disintegration. In particular, their attempt mainly consists of adopting and modifying business models into new horizontals, which requires a need for overcoming enormous amounts of inertia. This can happen through a concious opening-up of vertical *and* horizontal integration, such as in the case of δ_2, who integrates a broad spectrum of small players into a well-organized value network. The firm δ_7, who is bridging hybrid solutions between mobile handsets and PCs, thereby had less trouble in adopting their product roadmap to seemlessly work in both previously established parts of the value chain. And even the firm δ_3, whose major parts of firm identity and brand value consist of the tightly vertically integrated structures, has started to open-up that model, and allows competing operating systems to run on its products, suddenly maintains a dual relationship of core processor technology suppliers, and increasingly adopts standard technologies.

After all, it is the cost of vertical structures that can be seen as mainly responsible for this trend—the cost of retaining vertical integration in a broader industry is much higher (figure 3.9). As the value generation mechanism is increasingly dissaggregated and distributed, full-scaled vertical integration is no longer a viable solution, as it is costly to entirely recreate vertical structures. Hence, the cost of owning the entire set of previously distinct technologies exceeds the cost of owning a share of the converged technology domain:

$$C(\tau_{K_{1,2,\ldots,n}}) < \sum_{i=1}^{n} C(\tau_{K_i}), \quad \text{with } \tau_{K_{1,2,\ldots,n}} \equiv \cup_{i\in\{1,\ldots,n\}} \tau_{K_i} \text{ (cf. eq. 3.4)} \tag{3.6}$$

This leads to the occurrence of horizontal models, the emergence of platforms, and modularization (section 4.1). Any new form of verticalization, on top of platforms, is hence going below initial level.[23] Based on these observations, it is proposed as follows:

Proposition 5. (P_5) *Along the process of convergence, vertically integrated structures of value creation disaggregate into fragmented horizontal value creation systems.*

However, the evolutionary effects of disintegration and disaggregation, i.e., how firms react to this trend, can be seen in formation of

[23] These observations related to vertical disintegration suggest consensus with previous conceptions in literature, e.g., the propositions by Pavitt (2004) as summarized in table 2.5 (chapter 2).

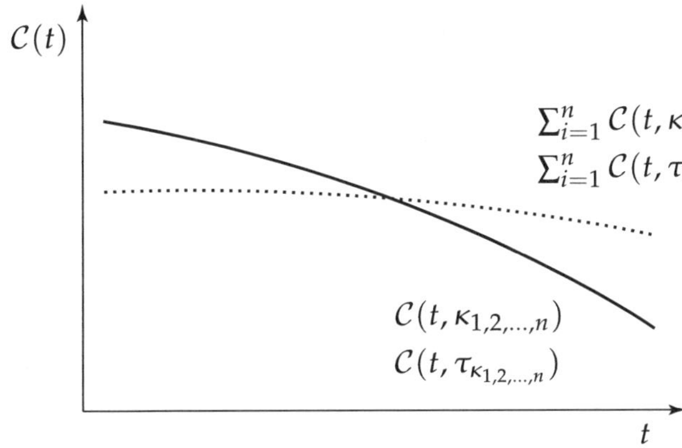

Fig. 3.9: Inflection in cost trajectories

new structures, strategic shifts, since retaining vertical integration is no longer a viable solution for them. Thus, it is inferred that convergence leads to a sudden structural entropy within previously organized and linear value generation mechanisms, resulting in a nonlinearization of such structures.

3.5.2 Adapted mechanisms of vertical reorientation

The previous section deals with selection in form of disaggregation and erosion of established vertical structures. Whereas these mechanisms are primarily based on a horizontal impetus, i.e., entry and new combination, as well as diversification (although partly lateral as well), this section will in contrast highlight the—*per se* somewhat paradoxical—new forms of vertical impetus along the—deconstructed—value chain. After all, literature suggests that vertical disintegration is followed by succeeding eventual reintegration (cf. Cacciatori and Jacobides, 2005; Fine, 1998).

In particular, although retainment of vertically integrated structures from firms' perspectives are no longer viable solutions, and horizontal orientations of firms allow more complex, and distributed forms of value generation mechanisms to emerge, certain trends of verticalization on an industry level do however continue. This implies firms to balance between horizontal strategies with potential for specialization, which, in particular, can not only go downstream, but also

Table 3.6: Deconstruction of vertically integrated structures

	Stage of convergence		
	Knowledge / *Technological*		*Applicational* / *Industrial*
Industry structure	Vertically integrated	Vertical disintegration	Horizontally integrated
Value generation structure	value chain		value network
Rule-changing process	deconstruction of closed structures		reorganization into open structures
Margin proprietorship	protected		shared
Cost of merging knowledge bases	$\mathcal{C}(\kappa_{1,2,\ldots,n}) > \sum_{i=1}^{n} \mathcal{C}(\kappa_i)$		$\mathcal{C}(\kappa_{1,2,\ldots,n}) < \sum_{i=1}^{n} \mathcal{C}(\kappa_i)$
Cost of diversifying technologies	$\mathcal{C}(\tau_{\kappa_{1,2,\ldots,n}}) > \sum_{i=1}^{n} \mathcal{C}(\tau_{\kappa_i})$		$\mathcal{C}(\tau_{\kappa_{1,2,\ldots,n}}) < \sum_{i=1}^{n} \mathcal{C}(\tau_{\kappa_i})$
Cost consequence	separate		integrate
Horizontal rationale	focus		diversify
Vertical rationale	diversify		focus
ENTRANT FIRMS	PIONEERING DISRUPTORS		VERTICAL ATTACKERS
Horizontal impetus	*new combination;* introduce new technologies between verticals		*infringe margins;* substitute business models through new horizontals
Example cases	α_5, α_3		β_4, β_3, β_5, β_6
ESTABLISHED FIRMS	PLATFORM CONSOLIDATORS		REDEFINING GIANTS
Horizontal impetus	*spill-over;* transfer and reapply technologies between verticals		*regain margins;* adopt and modify business models into horizontals
Example cases	γ_1, γ_5, γ_4		δ_2, δ_7

upstream along the value chain. Hence, one can observe companies who traditionally are located upstream in the value chain, such as carrier firms δ_1 or δ_9 to increasingly orient themselves downstream, by entering service domains, such as e.g., messaging solutions. Also, the firm δ_3 has significantly shifted downstream by entering the online music business. In parallel, firms such as α_3 or β_3 can be found as climbing upstream the value chain, attacking established communication services, and infringing incumbents' margins.

> "Interestingly, we launched [a product], which is a horizontal that can be used to create vertical views of content it stores. It is horizontal

in that any structured data can go into it, but it lets us create vertical interest areas by integrating its results into web search with a drop down form for certain vertical queries such as automobile, job, travel, or real estate searches. [...] again coming back to our core search we will build that core broad horizontal search shop and in some cases we need to get features of that which are related to verticals. Then we will do some of that." (informant 27, case α_3)

Similarly, firm β_6 experienced a shift away from mass and consumer markets towards enterprise customers, implying a need for the firm to refocus the efforts and to understand other layers of value generation. Also, the firm β_4 is currently undergoing a process of verticalization, which among others is being manifested in the shift from wholesale to direct selling.

Providing a slightly different perspective, the firm β_1 can be regarded as an entrepreneurial phenomenon of emerging verticalization within the carrier market. The firm is founded based on the belief in the solution's ability "to do a number of things to improve the underlying economics" (informant 15). At the current state, the market is highly ARPU-driven[24], with a constant strive towards minimizing churn.[25] Whereas launching new applications for potentially increasing ARPU still is a rather realistic approach, the problem occurs with the latter one. As the offerings by carriers are very generic, and thereby very similar, the market results in a constant price competition, which leads to heavy churn. In order to address this problem, the firm β_1 has the following approach:

"If you look at those segments, they are a mix: thousands of thousands of microsegments, with a lot of individual interests. [...] Practically, the easiest way to understand that is: look at the affinity credit card market—a huge trillion dollar industry—which started with airlines issuing credit cards, now practically everyone issues their own credit cards. When you get your bill, there are nowadays three parties involved: (1) the affinity organization, who can be everything, e.g., sports organization, church, charity organization etc.; (2) the credit card issuer, e.g., Bank of America, Citicorp, etc.; and finally (3) the credit card company, e.g., Visa, or Mastercard" (informant 15, case β_1)

Analogously to the affinity credit card model, the firm β_1 is working on a model to allow carriers to launch 'virtual' sub-operators

[24] Average revenue per user (ARPU)

[25] The terms *churn* denotes that subscribing customers often and broadly change carriers, i.e., high subscriber fluctuation from the perspective of a carrier.

branded through affinity organizations, thereby opening-up a realm of a practically unlimited amount of vertical markets, allowing traditional carrier companies to move away from the utility model, previously selling one-size-fits all, with same price for every one.

Also, the firm γ_4 has been "doing that since a couple of years" (informant 19). The firm has succesively moved from product sales to solution sales, and is currently working on orientating strategic activities towards the step thereafter.

> "What next? We need to continually change, its about making the customer experience work. First there was product knowledge, then service knowledge. [...] But we will also need to understand their [i.e., the customers'] business, that's were verticalization comes into play. [...] We have to look what the ecosystem looks like—like healthcare etc. [...]
> We're also, though, looking to make it increasingly relevant, so there is more and more vertical initiatives at γ_4. And those vertical initiatives, I think that, it's very hard to go from a box selling sales force, to a solution selling." (informant 19, case γ_4)

In other words, although convergence can erode entire previously established vertical structures and turn the industry into horizontal layers, later stages of the convergence process may still bring along tendencies of verticalization, which in the new configuration have to take place according to the new rules of the industry. Hence, firms tend to verticalize, but may not necessariliy strive towards vertically integrated structures anymore (table 3.7).[26] Based on these insights, with reference to proposition 5, it is suggested as follows:

Proposition 6. (P_6) *Along the process of convergence, collaborative dynamics of specialization and verticalization emerge, in response to fragmented horizontal value creation structures as suggested in* P_5.

The new specialization still has to happen in a verticalized environment, but cannot be implemented alone anymore, as it is too costly to regain vertically integrated structures (see previous section). Hence, partnering and collaboration seem to come into play even further.

> "That whole concept is new, about opening up the ecosystem, welcoming direct participation of partners, forging very strong relationships, navigating the difficulties of coopetition, where you cooperate in some times and compete in some other times." (informant 23, case γ_2)

[26] Similar observations were made in the work of Vesa (2006).

Table 3.7: Dynamics of specialization and re-verticalization

	Stage of convergence	
	Knowledge *Technological*	*Applicational* *Industrial*
Value generation mechanism	deconstructing	reconstructing
Firms' exogenous response	disintegrating	reintegrating
Structural shape	linear	nonlinear
Vertical paradigm	proprietary	collaborative
Vertical level of value generation	limited	expanding
Modularity of industry	increasing	decreasing
ENTRANT FIRMS	PIONEERING DISRUPTORS	VERTICAL ATTACKERS
Vertical impetus	disrupt and infringe upstream; explore downstream	
Example cases	α_3	$\beta_6, \beta_4, \beta_3, \beta_1$
ESTABLISHED FIRMS	PLATFORM CONSOLIDATORS	REDEFINING GIANTS
Vertical impetus	battle upstream obsolescence; specialize downstream	
Example cases	γ_4, γ_2	$\delta_9, \delta_1, \delta_3$

3.5.3 Commoditization and constricted economic incentives

In parallel with the previously elaborated trend of overall verticalizing structure of industry, the economic scope of actors in converging environments seems to narrow-down. In many cases, this trend can be motivated by increasing commoditization of upstream layers in respective underlying value chains, e.g., IT hardware, telecom equipment etc. Generally, this may cause established firms to struggle:

> "Telecommunications equipment is potentially the worst investment in the world right now, because of commoditization and give-away of intellectual property to silicon manufacturers, [...] there's no verifiable return on investment. We will see a lot of disruptions here in the future, [among] all of them." (informant 26, case γ_3)

Hence, the trend of moving downstream towards more service-oriented value generation levels goes in parallel with the fact that commoditized business areas do not leave room for margin development. As in the case of firm γ_5, the traditional business, namely building chipsets for handsets, is increasingly being complemented by new forms of distinctive value propositions. Hence, the firm provides a toolkit (section 3.2.1), allowing a value proposition beyond the technical platform.

> "If they do it their way, we become commoditized—just a dumb modem chip—and they do the money. [...] and Microsoft wants a world, where all money is made on the software side." (informant 28, case γ_5)

Again, carriers, such as δ_9 or δ_1 also represent an interesting example for this trend. Being attacked from peers at similar depths of value generation in information technology and consumer computing industries, they suffer from increasing obsolescence, which is based on their increasing commoditization. As in parallel, other players develop upstream, thereby entering the emerging vacuum, they become attacked from several directions at once:

> "As carriers get more and more commoditized, you just start to just be basically charging a flat fee for 'all you can eat' bit usage. That's not a that attractive model to them, so they may want a α_3 or β_3 to [become] the 2006 equivalent of railways—'you use my tracks, I want to charge you a fee to run your trains on my tracks'—and for a [firm like] α_3 or β_3, they hate that." (informant 25, case γ_1)

In particular, one can argue that information technology trajectories have reached a stage of evolution, which other segments such as phone or even electricity have some time ago, namely the progression into an utility model (cf. Carr, 2003). Hence, carriers with telecom legacy, who initially where able to create value-add through the convergence with IT, are now increasingly becoming commoditized on both underlying trajectories. Referring to this trend, firms such as γ_2 or α_1 base today's value proposition in the moving away from product innovation, towards process innovation (Utterback, 1994; Utterback and Abernathy, 1975, see also table 3.10):

> "That level of technology is not the differentiator. It's the business process that the technology enables, which is the differentiator." (informant 23, case γ_2)

But also entrant firms experience that form of inflection. Going back to the case β_6, the firm did not only move away from building all alone towards partnering with enterprises. This shift was accompanied by the change of business focus from business-to-consumer toward business-to-business. The firm α_3 provides an example for still being in scale phase, still building up solutions on targeting mass market without any specific limitations, although the verticalization and stepwise emergence of vertical compementary applications—i.e., the scope phase—can be anticipated.

Table 3.8: Economic inflection from scale to scope

	Stage of convergence	
	Knowledge — *Technological*	*Applicational* — *Industrial*
Economic rationale	scale	scope
Market focus	mass	niche
Industry dynamics	differentiation; technological intersection; uniqueness	commoditization; business model collision; redundancy
ENTRANT FIRMS	PIONEERING DISRUPTORS	VERTICAL ATTACKERS
Manifestation of change	scope substitutes scale	
Example cases	α_3	β_6
ESTABLISHED FIRMS	PLATFORM CONSOLIDATORS	REDEFINING GIANTS
Manifestation of change	scope complements scale	
Example cases	γ_4, γ_5	δ_9, δ_1

In the face of commoditization of underlying technologies, horizontal differentiation becomes less substantial, further increasing the emphasis on verticalizing strategic approaches (table 3.8). It is proposed as follows:

Proposition 7. (P_7) *Along the process of convergence, the value creation opportunities for intersectional applications develop from mass to niche markets.*

As with the handset devices by firm δ_7, many other firms have to go away from being too multifunctional, integrating everything and too large, into development approaches of scope and niche. In this context, comparing a certain product category of competing device vendors such as δ_2 or δ_6 with those of his company's product offering, informant 41 concluded:

> "It's a freakin' Frankenstein! It can do a lot of things, but nothing really well." (informant 41, case δ_7)

3.5.4 Competitive dynamics and convergent dominant designs

Finally, a few general observations regarding the development of competitive dynamics over time are introduced. Similar to any major technology life cycle as induced by a discontinuous event (Abernathy and

Utterback, 1978; Anderson and Tushman, 1990; Tushman and Anderson, 1986; Utterback, 1994; Utterback and Abernathy, 1975), the general degree of competition, i.e., the amount of either technologies, products, solutions or entire companies coexisting at a certain point of time, seems to relate to the respective stage within the convergence process. In particular, the discussion with informants in general reflected a general perception of earlier stages of convergence being characterized by increasing competition, whereas at some stage, at latter stages of convergence, the competition starts to decline.

In this context, the overall competition was perceived as switching in its form as well. During an increasing number of competitive players, the major sources of competition were described as inertia, substitutes and direct competition. On the other hand, during latter stages, i.e., decreasing competitive pressure, the source of competition was perceived as consisting mainly of direct competition. This may be seen as an indicator for industrial convergence having taken place, as the formerly substitute or inertial competitive sources either have dropped out of the race or become direct, as the industry boundaries evolve.

For instance in the firm β_6, at a certain point of time, competitors dropped out or refocused, and today the company finds itself basically alone in their strategic approach. In the case of firm β_5 there are about a handful of competing providers out there (among them β_2), but a high degree of consolidation is going on currently. However, in the current stage of the convergence process, it is perceived as rather impossible for any new firms to enter this industry and build up all the required capabilities from scratch. A more realistic new competition could be seen in application providers extending their applications into mobile ones, which still would result in a too complex multitude of systems. Among these firms, the dominant design, i.e., mobile push email with connectors to established back-end infrastructures, seems emerged as a *de facto* standard.

Hence, it is postulated that similarly to traditional cycles of industrial change, the convergence process may at some stage result in a *convergent dominant design*, which is likely to emerge somewhere between technological and applicational phases of the process (table 3.9). This is particularly the case when the growth of opportunities based on the intersection of knowledge bases and technological regimes stabilizes. The difference from the tradional conception of a dominant design[27] is, that the converging dominant design spans previously

[27] Cf. Anderson and Tushman (1990); Suarez and Utterback (1995); Utterback (1994)

Table 3.9: Overall changes in competitive landscape

	Stage of convergence			
	Knowledge	*Technological*	*Applicational*	*Industrial*
Competition and inter-firm rivalry	increasing		decreasing	
Major sources	inertia; direct; substitutes		direct	
Technological maturity	era of ferment		era of incremental change	
Variance of designs	divergent competing designs		convergent dominant design	

distinct industries, thereby terminates any major design variance in underlying industries.

Furthermore, the real implications of the emerging convergent dominant design begin to substantiate at latter phases of convergence. As soon as convergence proceeds throughout commercial stages, and the industrial convergence starts to be complete, the starting consolidation takes place along the new industry rules. In particular, this may imply that the consolidation takes place beyond the initial level, i.e., the baseline of number of technologies, products, solutions or firms in the previously distinct industry configurations. In other words, this new consolidation process may render less firms than before (figure 3.10).

Further stretching the concept of dominant design, one can in this new form of consolidation distinguish between two forms (figure 3.10):

Definition 12. (Local dominant design) *When the amount of technologies, products, solutions or firms starts to decrease at intermediate stages of the convergence process, a local dominant design emerges. At this point, the initial variation, implied by the wave of new intersectional opportunities starts to decline.*

Definition 13. (Global dominant design) *When the convergence process reaches mature stages, and the initial variation, implied by the wave of new intersectional opportunities, has significantly declined, global dominant designs may occur. At this point, an inter-industry consolidation is initiated, eliminating technologies, products, solutions or entire firms, which were previously mutually unrelated.*

For instance, by pushing the development of search technologies, firms α_3 and β_3, are currently working towards a local dominant design, outperforming a variety of upstart and established competitor firms within that particular field. But extrapolating into the future, the path might eventually lead to a global dominant design, as the firms not only infringe previously unrelated business areas (such as e.g., online bookstores), but start to outperform them. Also, β_2 has previously been dealing with a local dominant design, which occurred among mobile e-mail solution vendors. However, the emerging global dominant design can be seen in higher levels of messaging services in general, leading to a consolidation among both entrants and established firms (which lead to the acquisition of the company by δ_2). Firm δ_2, in turn, starts to compete with established players from the camera industry as well, and the local dominant design is still to be seen. It is probably only a question of time, before first consolidation activities between camera and mobile handset industries start to emerge. Also, the firm δ_2 recently merged a division of the company with that of a major competitor. Although in this case both players originally find themselves rather on the telecommunications side of convergence, they do however still carry very different corporate legacies.

By selling a subsidiary of the firm to a major player in the movie and entertainment industry, firm δ_3 shows strong trajectories of positioning itself within that field of business. Finally, Internet service-oriented firms, such as firm β_3 increasingly collaborate with carriers such as firm δ_1 or δ_9, and related manifestations of global dominant designs will be visible in the upcoming takeovers in this field. In summary, the following proposition is formulated:

Proposition 8. (P_8) *The process of convergence undergoes a cycle of successive increasing and decreasing competition, which is successively punctuated by the emergence of local and eventually global dominant designs.*

3.6 An integrated framework of evolutionary dynamics

3.6.1 Evolutionary effects within the dichotomy

> "When these things come together, there's a real opportunity for disruption." (informant 17, case α_5)

As initially indicated in section 2.3.3, convergence phenomenon implies an impetus within the dyad of innovation dynamics. Previously distinct, and *per se* incremental, trajectories of technological

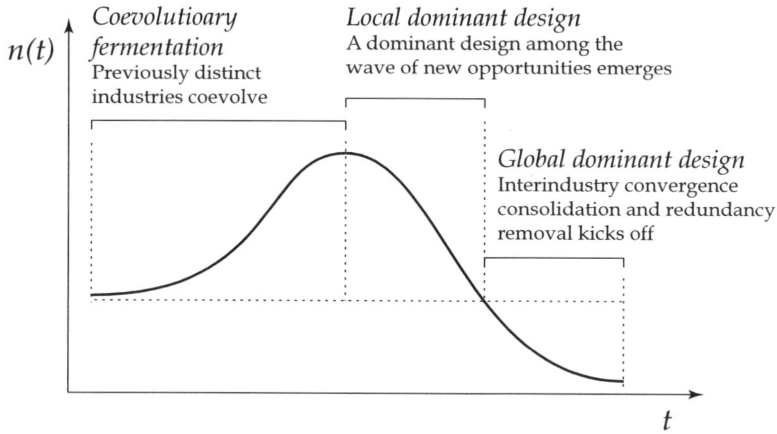

Fig. 3.10: Local versus global dominant design

change can in their confluence render disruptive effects and ignite rule-changing processes. But, this does not only bring along a dyadic shift in terms of innovation dynamics. In particular, the type of innovation seems to change as well, i.e., the convergence effect does not only punctuate incremental, stable states and turn them into discontinuous disequilibria. The resulting disruption ignites an entirely new—yet coevolutionary—innovation life cycle.

The evolutionary perspective, as illustrated by the process of cascading transitions (section 3.2), encapsulates a rule-changing process from knowledge and technology to application and industry. Observations of firms along this commercializing process of convergence suggest a certain conformity with the traditional dynamic model on innovation patterns as introduced by Utterback and Abernathy (1975). As the convergence process matures, firms increasingly are dealing with process innovations, instead of product innovations. In other words, a congruence between convergence stages, and innovation patterns is inferred (table 3.10). Hence, the observations on effects of convergence serve to disentangle the convergence phenomenon from an evolutionary perspective, which in the following section is presented as an integrated model of congruent change mechanisms, all of which in their confluence are causing changes in innovation dynamics.

Table 3.10: Innovation typology and stage of convergence

	Stage of convergence			
	Knowledge	*Technological*	*Applicational*	*Industrial*
Pattern/type of innovation	product		process	

3.6.2 Juxtaposition of observed conceptions

A combination of observations made in sections 3.4 and 3.5 is used for a inducing a multilevel framework of phenomenon. In assuming timely congruency, it must not be taken for granted that all observed inflection mechanisms happen at a timely synchronized framework. However, as the convergence process is used as the underlying reference framework, the timely congruence of inflection events does not necessarily relate to timely interdependence, but to the fact that they were observed in the same firms, which in turn were located in the respective same phases. A summarizing overview of observations in table 3.1 to table 3.10 yield the juxtaposition in table 3.11.

Hence, the convergence phenomenon represents a particular form of discontinuous change. Strategic inflection points can be widely observed, both within and between firms. On a firm level, search activities change from closed, uni-disciplinary into open, multidisciplinary innovation structures, and depending on the intertial consequences of the transition, firms may induce competence-destruction and co-evolutionary lock-in to themselves. On an industry level, established industry boundaries may increasingly become blurred, as a consequence of horizontal spill-over and disaggregation of established vertical structures. After all, convergence seems to lead to certain consolidation mechanisms over time, when new dominance situations start to emerge, the competition decreases and the variation experiences significant selection. These observations combined with the theories of Tushman and Rosenkopf (1992); Utterback (1994); Utterback and Abernathy (1975) render an integrated overview in figure 3.11.

3.7 From challenge to practice

Based on these findings, convergence seems to represent a metamorphosis from two or more previously distinct indutrial domains into a new, conglomerate area, which from a managerial perspective represents more than solely the sum of the parts. For leveraging convergent

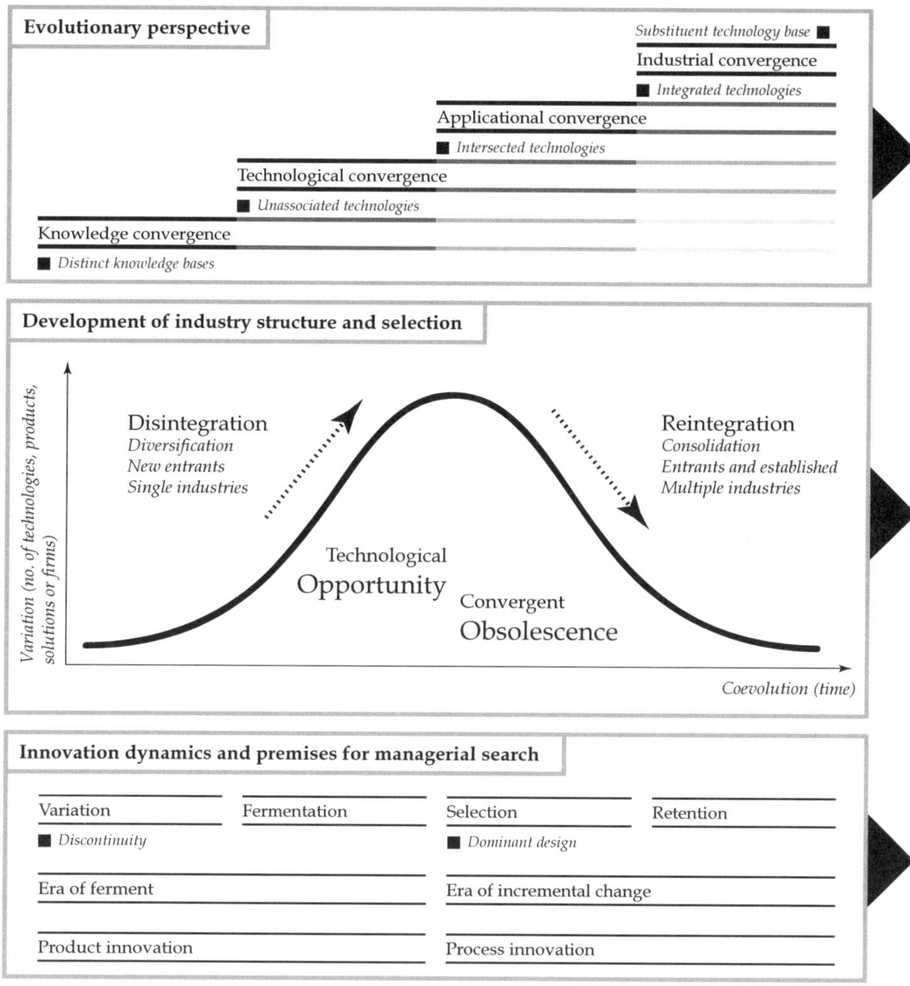

Fig. 3.11: Overlap of different models

synergy, and managing growth, it is argued that distinctive understandings of the evolutionary implications of convergence phenomena are needed on a strategic level.

In particular, the convergence phenomenon turns out to represent an autonomous movement, which can be characterized by serendipity, and cannot be directly associated to firms´ actions (a phenomenon between search and selection). Furthermore, convergence can be understood as the development of underlying trajectories, i.e., convergence does *not* necessarily take place when firms bundle components into comglomerate products and solutions, it is when underlying bases of knowledge and technologies eventually come together.

This leads to the conceptual proposition, that one cannot—and should not—attempt to manage *the* convergence, but rather manage *through* convergence. Accordingly, the following two chapters will deal with the question of how to manage through this specific form of technological change. From the same data set as introduced in this chapter, managerial responses to the observed phenomena will be analyzed and discussed. Continuing at the understanding of strategic inflection points, chapters 4 and 5 will deal with the managerial challenge of mastering the switch, by making use of current capabilities, thereby still resulting in something completely different than before. Whereas this chapter dealt with mainly distinguishing between endogenous and exogenous effects of convergence, the strategic approaches in the following chapters will be based on aligning endogenous management activities with exogenous dynamics.

Table 3.11: Overview of observations

	Stage of convergence	
	Knowledge *Technological*	*Applicational* *Industrial*
Pattern/type of innovation	product	process
Focus of rule change	exploration	exploitation
ENDOGENOUS CHANGE		
Local knowledge bases	single	diverging
Competence challenge	opportunity; novelty	inertia; lock-in
Control of innovation process	proprietary	open
Scope of innovation activity	internal	external
Dynamics of innovation activity	in-house	out-house
Disciplinary search processes	introverted	extroverted
Disciplinary task assignment metrics	static	dynamic
Base for expertise	within disciplines	between disciplines
Role of geographical clusters	protected	disentangled
EXOGENOUS CHANGE		
Global knowledge bases	multiple	converging
Value generation structure	value chain	value network
Margin proprietorship	protected	shared
Horizontal rationale	focus	diversify
Vertical rationale	diversify	focus
Value generation mechanism	deconstructing	reconstructing
Firms' exogenous response	disintegrating	reintegrating
Structural shape	linear	nonlinear
Vertical paradigm	proprietary	collaborative
Vertical level of value generation	limited	expanding
Modularity of industry	increasing	decreasing
Economic rationale	scale	scope
Market focus	mass	niche
Industry dynamics	differentiation	commoditization
Competition and inter-firm rivalry	increasing	decreasing
Major sources	inertia; direct; substitutes	direct
Technological maturity	era of ferment	era of incremental change
Variance of designs	divergent competing designs	convergent dominant design

4

Capabilities for coevolutionary contingency

As elaborated in the previous chapter, the convergence phenomenon implies changing characteristics on innovation trajectories, both on industry and firm-level. In response to these temporal dynamics, the development and measurement of managerial approaches needs to take place under the evolutionary premise. Whereas the previous chapter primarily focused on the perspective of economic selection, i.e., the evolutionary perspective, which is mainly characterized by innovation dynamics, and exogeneous effects, this chapter will further emphasize on firm-level activities, i.e., search (defition 1). Based on the previously identified challenges related to respective phases and competitive constellations within the convergence process, this chapter will identify inherent firm-level capabilities for managing the dynamics of convergence. These capabilities are based on analyzing how firms from previously introduced case set responded to their respective challenges and changing conditions. Based on a classification of these capabilities, it is further attempted to develop guidelines for managing innovation through evolutionary convergence cycles, which serve as a basis for an induced framework for capability development.

In the study of firms within the ICT set (table A.1, p. 210), distinct phases of the convergence process have been identified and described, which can be associated with specific evolutionary and strategic challenges (section 3.2–3.5). It was additionally observed that not only does the phase within the convergence process matter. Firm-specific legacies play a crucial role in determining strategic opportunities, as firms differ in size and history, and accordingly have different structural inertia (cf. Hannan and Freeman, 1984). Actors in intersectional environments have to navigate the convergence process within the realm of two dimensions: on the one hand, they need to follow endogenous (i.e., firm-level) developments, and on the other hand, they have to respond to exogenous (i.e., industry-level) evolutionary changes (which in turn might alter endogenous development paths). This realm of evolving and reacting results in the notion of coevolution, which formed the conceptual basis for distinguishing between different strategic constellations, and for identifying classes (section 3.3.1, definition 7).[1]

Accordingly, the development of managerial recommendations has to differentiate between specific coevolutionary constellations. Allowing further identification and articulation of managerial capabilities for dynamic adaptation and renewal, it is hence assumed that there is not *one* best way to manage convergence under the variety of different strategic configurations, which in this study are represented through the coevolutionary classes. Based on this assumption, a contingency perspective is selected for examining capabilities among the case firms.

There exists a broad variety of conceptions and interpretations of contingency theories related to strategy and organization, based on classic studies of Burns and Stalker (1994), Woodward (1965), Lawrence and Lorsch (1967) as well as Chandler (1962). The common base line among these studies consists of the proposition that firm performance is a consequence of the fit between several factors, such as structure, people, technology, strategy and culture (Tosi and Slocum, 1984). In particular, according to the seminal work by Lawrence and Lorsch (1967), firm performance can be viewed as a function of organization and environment:

> "The environment with which a major department engages is decided by the key strategic choice, 'What business are we in?' Once that decision is made, whether explicitly or implicitly, the attributes of the chosen environment can be analyzed. [...] Internal attributes of the organization, in terms of structure and orientation, can be tested for goodness of fit with the various environmental variables and the predispositions of members. Unit performance (which will have to be judged by a number of dimensions, of which profitability is only one) emerges as a function of this fit." (Lawrence and Lorsch, 1967, p. 209)

[1] Kauffman (1993) argues that *all* evolution is really coevolution: "The true and stunning success of biology reflects the fact that organisms do not merely evolve, they *coevolve* both with other organisms and with a changing abiotic environment." (Kauffman, 1993, p. 237)

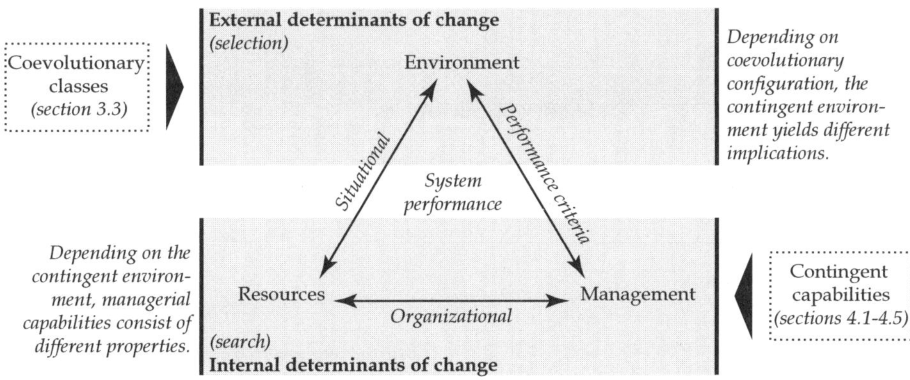

Fig. 4.1: Determinants and relationships in a contingency model

Aiming at a rather integrative approach with their work on a general contingency theory of management, Luthans and Stewart (1977) discuss the notion of system performance within the dimensions of environment, management as well as resources. In general, contingency theories build on the argument of the right choice of strategic action being contingent on the specific circumstances of the organization, i.e., there is no universal strategy that works well for every organizational context.[2] Combining this approach of managerial model construction with the resource-based view (RBV) of the firm (Wernerfelt, 1984), the concept of contingent RBV (Brush and Artz, 1999) provides a logic for formulating firm strategies: aligning both tangible and intangible assets, i.e., both resources and competences to fit the environment. Building on the model of Luthans and Stewart (1977), figure 4.1 depicts the determinants and their relationships in a managerial contingency model. In particular, the model distinguishes between both external and internal context, as a contingency-oriented framework needs to take into consideration the environmental characteristics of the industry, the resource configuration of a firm, as well as required managerial determinants.

[2] Hofer (1975) positions the development of contingency theories as a feasible approach for business and corporate strategy, "unless one is willing to admit the possibility that there exists some strategy or set of strategies which are optimal for all businesses (corporations) no matter what their resources and no matter what environmental circumstances they face—an assumption that is inconsistent with all research studies on business (corporate) strategy conducted to date—any theory of business (corporate) strategy must be a contingency theory." (p. 785-786)

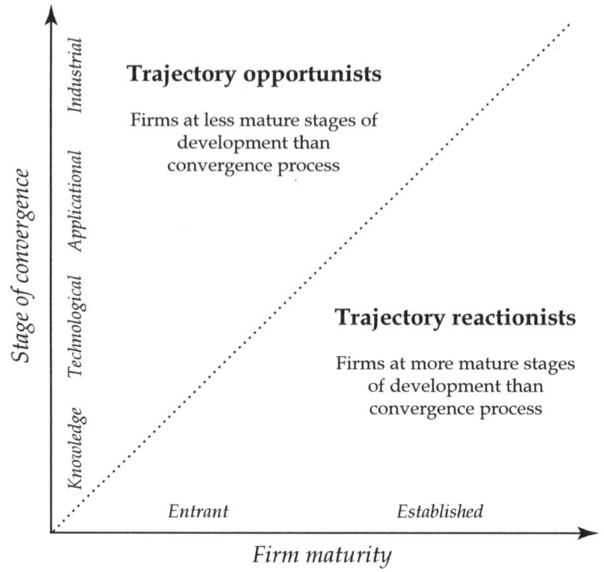

Fig. 4.2: Leading vs. following the convergence trajectory

In defining the context of the contingent environment, the distinction provided by the coevolutionary classes is further adopted, consisting of the dimensions of stage and maturity. In a simplified view, the contingent environment of a firm within a coevolutionary class is determined by the the remainder of coevolutionary classes.[3] Generally speaking, one can within that framework distinguish between constellations where firm-level search activities happen either as an action, or as a reaction to the trajectory of development. Based on the same frame as used for distinguishing coevolutionary classes within, the following additional definitions are presented:

[3] Speaking in symbolic terms, let $\{C_1, \ldots, C_n\}$ represents a set of coevolutionary classes with specific firms $f_{1,i}, \ldots, f_{m,i} \in C_i$. Then, the contingent environment of a given firm $f_{j,i}$ is determined through $\{C_1, \ldots, C_n\} \backslash C_i$. For instance, the contingent environment of firm δ_1, which belongs to the class of reincarnating giants, is determined by the classes of pioneering disruptors, vertical attackers and platform consolidators, as these classes coevolve with firm δ_1 and contribute to the exogenous selection mechanism that affects the individual firm's conditions for search.

Definition 14. (Trajectory opportunists) *Firms at less mature stages of development than the convergence process can be viewed as trajectory opportunists.*

They are in their organizational and strategic age ahead of the convergence process, they can move more easily, and might have an affinity of being able to explore emerging potentials of convergence in a better way.

Definition 15. (Trajectory reactionists) *Firms which themselves are at more mature stages of development than the convergence process are denoted as trajectory reactionists.*

These firms have a tendency to react to convergent changes in a follower-way, carry the burden of tradition, and have to overcome structural inertia (figure 4.2 and definitions 14–15).

Hence, these coevolutionary constellations, occurring within the realm of leading or following the convergence trajectory, represent the basis for the environmental context of the contingency model (figure 4.1). As seen throughout the previous chapter, depending on the coevolutionary configuration, the environment yields different implications to firms. As this chapter will elaborate on the context of managerial search, the external context will be taken for granted, thereby acting an independent variable. Instead, the firms from the ICT case sample will be examined with a lens on the internal determinants of change, aiming at identifying capabilities between management and resources. Hence, along with this dimension, firm-level capabilities will be elaborated and distinguished in contingency to the previous positioning within the four coevolutionary classes (figure 3.4). Figure 4.3 provides an initial overview of identified managerial capabilities, which will be described in the following sections (4.1–4.5).

4.1 Aligning horizontally for developing vertically

As the process of convergence deconstructs established value chains on the one hand, the industrial transformation creates opportunities for horizontal business models (section 3.5.1). On the other hand, however, as new forms of vertical orientation and specialization are needed (section 3.5.2), firms need to be able to leverage both the breadth of converging knowledge bases, as well as the response to specific customer needs, focusing on specialized vertical segments. On the contrary, building solely on vertical business models might

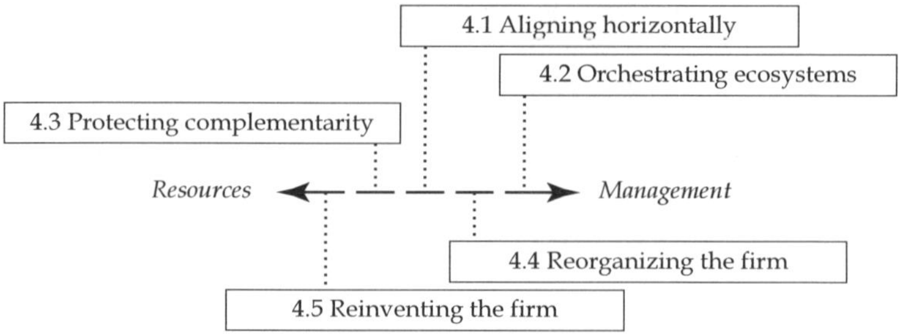

Fig. 4.3: Capabilities between resources and management

cause firms to end up in highly path dependent positions, from where it becomes difficult to move horizontally towards new spill-overs of knowledge bases.

In such a situation, the platformization concept was observed as a feasible strategic approach for capturing convergence opportunities and mitigating path dependency. In literally building the company around a platform,[4] which spans the converging set of previously distinct knowledge bases and technologies, firms develop capabilities for leveraging the synergetic effects on the one hand, allowing complementary innovation in the form of vertical products, solutions and even entire business models to be built on top of that (cf. Cusumano and Gawer, 2002; Gawer and Cusumano, 2002). Hence, the organization of firm capabilities as a platform, i.e., consisting of convergence-spanning assets such as knowledge, technology, organization and partnerships, allows the firm to minimize lock-in effects and structural inertia, and provides a growth path into—yet unknown—new cycles of convergence. On the other hand, being successful in leading a platform within the coevolutionary environment may entail other firms to gravitate around the business model, allowing further opportunities for inducing resource dependency and coevolutionary (cf. sections 4.2 and 4.3).

Based on the observations as elaborated in the following sections (4.1.1–4.1.4), a set of capability elements is identified in contingency

[4] If other firms invest specific assets to preserve compatibility with a given firm's product, even as that product evolves, then that product can be called a platform (Gawer, 2000).

to the coevolutionary classes. These elements are summarized in table 4.1, and yield the following proposition:[5]

Proposition 9. (P_9) *Contingent on its coevolutionary class, a firm needs to develop a specific dynamic capability for understanding, developing and maintaining a platform advantage for capturing the horizontal spill-over between knowledge bases and its higher-level applications.*

Table 4.1: Capability elements for developing platform advantage

Pioneering disruptors	*Vertical attackers*
explore new technologies for potential platform construction	exploit new technologies; leverage spill-overs between downstream layers
stick to open standards	identify existing technological and business model conflicts and construct platform based on that
publish and explore open platform, allow others to build on top of that	create growth path based on convergent platform
create clear differentiator for platform	
Platform consolidators	*Reincarnating giants*
explore complementary assets	exploit complementary assets
align existing technological assets into platform, spanning converging space	identify out-of-the-box platform opportunities among existing resources
acquire external competencies for creating new complementarities	acquire external competencies for upgrading and substituting internal competencies
open-up innovation process to smaller players	
develop total platform advantage	
capability platform thinking (technology, knowledge, partnerships, organization)	

[5] Needless to say, the occurrence of platforms does not necessarily imply the emergence of a convergence phenomenon. The concept of platforms *per se* may as well occur within industries with protected boundaries, e.g., with the aim of sustaining upward compatibility or reusing components. Instead, a managerial capability founded on platform thinking may provide a tool for creating a growth path through a convergence process, as the platform may capture the horizontal breadth of mutually approaching knowledge bases.

4.1.1 Pioneering disruptors: explore platforms

Being at early phases of the convergence process, where technology bases are not fully existent yet and still are within the fermentation phase, pioneering disruptors in general find it hard to commit too much resources on platform concepts, as they still are in the phase of exploring technologies rather than exploiting them.

In the case of α_1, "our approach is to stick to standards" (informant 30), working towards having systems easy migratable, allowing relatively agile rapid moves when needed. However, standards themselves were perceived as significantly behind vendor technologies, as people generally want to build something that is well beyond where standards are today. Still, from the business process engineering perspective of the company, the clear benefit of standards was perceived in their ability to accommodate the complexity of the variety of different existing processes, resulting in a platform, which will help the convergence.

Similarly, in the case of α_2, the strategic significance of platforms is not that tangible yet. Since the company "had to find something that scales" (informant 11), it was seen as an obvious and essential factor, but at this point it was not necessarily a critical differentiator yet, but "is rather on-off, you need a platform to scale business".

The firm α_3 is in the phase of shaping the platform around search technologies, and in turn building the entire company around that, allowing the firm to address "hundreds of little micro verticals" (informant 27) on top of that. Still, the current focus is on shaping the horizontal scope of the platform, with a strategy to only address a focused set of the vertical segments. Additionally, the firm follows an open approach to exploring new elements of the emerging platform. New components and features of the platform, which in many cases consist of rather prototypes and beta-versions, are launched in dense cycles, are published for free, allowing users to explore the technology and even building products on top of that.[6]

An informant from a competing firm commented on the innovation strategy of firm α_3:

[6] From the perspective of early stages of the convergence process, where a firm's explorative efforts strive towards developing platform concept for gaining horizontal foothold, the approach of letting users drive innovation may yield beneficial results, as they generally tend to apply higher levels of technological diversification (but lower levels of specialization) than manufacturer-driven innovation (Lettl, Rost, and von Wartburg, 2006).

> "[...] they simply publish the platform, so that other people can then innovate on top of the platform. So, that's a different model. In many ways, same approach as open source: build your own products on top of an open source, and then let the rest of the community in some sense, help build the other components that you cannot build yourself." (informant 16, case δ_6)

4.1.2 Vertical attackers: exploit business model conflicts

Entering increasingly commercialized stages of the convergence evolution, firms within the class of vertical attackers can disrupt established business models through integrating mutally competing pieces of existing business models into a platform. Hence, players within such a class of being at rather mature stages of the convergence process, but being themselves rather young, independent and agile organizations, need to develop capabilities for anticipating and identifying existing as well as emerging technological or commercial conflicts, and construct a platform based on bridging these gaps. Again, being a small firm, the growth path is in such a scenario based on building the company around the platform, with the additional advantage of benefiting from the fragmentation and heterogeneity of existing business regimes in the new convergent industry landscape.

Similarly to the firm α_3, the firm β_3 builds the company around a sophisticated platform for data search and categorization mechanisms. But compared to firm α_3, the firm β_3 finds itself in a more commercialized stage of the convergence process, where the return of vertically oriented structures can be envisioned, and where building vertical business models on top of the existing platform has become a core part of the business.

> "It is easy to use that service, to extend it, and to enable a new one. We can't build everything by ourselves. [...] If you can enable to become the platform behind it, you have a lot more reach, and over time, you build a really, really strong base for other businesses and other service providers providing additional value." (informant 8, case β_3)

Hence, the firm β_3 has developed platform capabilities as a facilitator for a verticalized market, either building the vertical components by themselves, or providing open innovation access for other firms to complement the platform towards niche segments.

The firm β_2 spent large R&D resources into the development of a smart platform, that mastered the technological conflicts of mobile data access. In particular, the firm decided not to go directly for the market pull, i.e., developing focused vertical solutions matching specific customer needs. Instead, the strategy was to invest a significant amount of resources into generic data synchronization mechanisms and technologies. This resulted in a strong platform and industry brand, allowing the firm to master the first generation of vertical demands, i.e., mobile and offline access to corporate informaton managagement data (such as contacts and calendar). Additionally, the resources invested into platform capabilities started to pay-off as new customer demands arose based on an increasing horizontal span of convergence. Emerging needs for mobile e-mail access could easily be implemented as a feature of the platform, not only technologically, but also in terms of organization, partnerships, and sales, as the company already knew the rules of the emerging—and converging—industry. This allowed the firm to show a clear growth path and tangible long-term product roadmap compared to its competitors, which in the end resulted in the firm being acquired at a high price by δ_2. And even the next wave of distributed and disconnected mobile data access might be significantly influenced by the dominant joint industry position of δ_2 and β_2, when smart synchronization protocols and interfaces will be needed for disconnected shared filesystems and applications around that.

Similarly, the firm β_6 experiences a clear growth path based on their "rock solid platform" (informant 10), which is build on open standards, such as e.g., defined by the World Wide Web Consortium (W3C). This lowers the firm's path dependency.

> "[...] we have built this with enough flexibility, that we want to anticipate the next technology trends, rather than be caught behind the 8-ball" (informant 10, case β_6)

Although still operating in stealth mode, the firm β_1 builds the company's value proposition around the platform, which mainly consists of a large and sophisticated database system, allowing "highly attractive new vertical applications on top of that" (informant 15). As the platform allows rich and detailed data on subscribed users' profiles and behavior, all marketing and segmentation activities are based on that engine. The firm leaves the platform open for other firms to innovate on top as well, providing an interface which allows the development of customized and tailored end-user offerings through third parties.

4.1.3 Platform consolidators: search for complementarities

Whereas entrant firms can develop internal capabilities independently from any inherited organizational or technological legacies, established companies should primarily rather attempt to recombine existing resources, or at least parts of them, into new business opportunities, before directly deciding to reinvent the entire firm (section 4.5). The idea of combining something new with something old can allow the company to exploit[7] the advantage of existing complementary assets towards the small firms who infringe the market with new innovations (cf. Katila and Ahuja, 2002).[8]

Accordingly, the key to convergence within the class of platform consolidators seems to lie in the ability to align internal and external technological competencies into a platform, which through confluence allows the firm to explore and shape the emerging multidisciplinary knowledge base. As the convergence process itself is at rather early stages in these cases, firms within this class can reallocate and acquire resources as well as reorganize the company to result in a capability platform, securing the future "total platform advantage" (informant 24) for any emerging vertical application.

Hence, instead of constructing a new platform completely from scratch, the firm γ_4 decided to leverage core assets of the firm into a platform strategy, aiming at developing and shaping assets which complement emerging convergent applications. In their case, building on the vast array of network equipment and related products, the firm's convergence capability centers around "using the network as a platform" (informant 19). Additionally, current services and applications will be aligned and positioned as a vertical extension on top of that platform.

---

[7] In a way, one could regard this strategy as the incumbent firm trying to take the role of the 'dilemma' in entrant firms' 'innovator's dilemma', i.e., outperforming entrants based on their own innovations (cf. Christensen, 1997).

[8] Based on a study on the typesetter industry, Tripsas (1997) similarly suggests that specialized complementary assets play a crucial role in buffering incumbents from the effects of competence destruction, whereas further commitments onto isolated technical capabilities has a tendency to lead to misleading results.

Taking the argument one step further, Rothaermel (2001) finds that from an incumbent's perspective within the biopharmaceutical industry, firms who focus their collaborative innovation strategy on exploiting complementary assets outperform those who focus on exploring new technologies. This suggests an approach of leveraging complementary assets to be more favourable than building new technological competencies.

"We will always be a networking company. I think that we are saying that the network is the *platform*, and what happened in voice [technologies] is becoming an offering all for that platform." (informant 19, case γ_4)

Based on this starting point, the firm is currently investing resources both into developing the platform horizontally, as well as into building vertical applications. After voice services have become fully commoditized on that platform, the company anticipates video as the next convergence cycle.

Likewise, the firm γ_1 is in the process of responding to the convergence transformation through aligning existing technologies and R&D projects into a horizontal baseline, allowing more cross-fertilization and synergy.

"It is not about how to merge all these radio technologies, but how to get them onto a platform, that yet additionally is open for new innovations to come down the pipeline. [...] an architecture, platform, that is not stuck and doesn't need reinvention. [...] Truly, a configurable platform, [...] and [we] may not provide all the components." (informant 24, case γ_1)

These efforts have so far resulted in a platform product, which broadly spans the emerging convergent space, and will position the company as a key supplier in many leaps of upcoming computing and communication hardware. Through consolidating and complementing existing assets internally, the company was able to create a strategic horizontal value proposition. But also the exploration of external complementary assets is part of the current path, as current innovation activities are about streamlining the firm into the "total platform advantage", based on which the firm will "rely on the ecosystem and partners quite a bit" (informant 24). This platform advantage will ensure the firm's distinctive capabilities against direct competitors, such as other process technology vendors, or vertical competition, such as cellphone manufacturers who aim towards developing proprietary integrated communication chipsets. Also, as the convergence process still is at a rather early and exploring phase, the firm was not directly forced to solely exploit complementary combination, but was also able to extend internal resources through acquiring capabilities which they did not have before (e.g., emulating analog modems on the CPU, i.e., digital communications protocol processing).

For the firm γ_2, the development of convergence-related capabilities represents a major strategic inflection, as the firm "hasn't before been a platform company either" (informant 23). As the technological foundation of the company however already mainly consisted of a horizontal product, it was rather the understanding and mindset on this issue, as well as the organization of associated processes, that had to change. As in the case of γ_2 the array of vertical segments based on the platform is extremely broad, fragmented and still specific, the firm decided to meet the megatrend of open innovation, and de facto make their customers develop the verticals by themselves (section 4.2.3). Based on such an approach, the firm invests its own development resources into the maintenance and horizontal growth of the platform, whereas the external verticals—the customers—provide the complementary assets, and sustain the growth path of the entire company.

4.1.4 Reincarnating giants: complement or acquire platform

In the class of reincarnating giants, a more severe sense of urgency related to responding to convergence challenges has to be created. As the convergence process has reached rather commercial stages at this point, there is not much time left to explore new technologies, as the market is running based on convergent business models—either with or without these established firms.

Particularly carrier firms tend to find themselves in an "ongoing identity crisis" (informant 15), as they have to search for a complementary match between core telecom infrastructure models, and reinventing themselves regarding the layer above. In emerging stages of the convergence process, incumbent telecom carriers had a general tendency to neglect the disruptive potential to their core business, and rather declared themselves as the clear winning stakeholders—as they own the infrastructure, they also own the business. In many cases, this turned out to be a misconception, which in addition to an internal reluctance to reinvent highly profitable established business models, resulted in today's situation, which can be characterized by structural, and other forms of inertia. With decreasing margins on traditional circuit-switched service models, many of these firms cannot just like that make a complete turnaround and commit onto applications (i.e., scope), while infrastructure and other assets are focused onto infrastructure (i.e., scale), such as e.g., controlling processes, billing mechanisms or marketing concepts. Among these, the firm δ_9 has the advantage of being a rather small-sized carrier compared to many of its global competitors. Still, the origins of the current path dependency are lying in the organizational inertia, as well as the corporate culture, with its long burden of tradition. Hence, before expecting any major success in reincarnating the company into exploiting radically new opportunities within the constantly advancing realm of convergent business models, the company has an internal project ongoing, with the aim of "changing the mindset" (informant 5). Based on the development of such a fundamental capability, the firm hopes to gain more anticipative capabilities and internal entrepreneurship. Based on this foundation, the next upcoming step to minimize the current lock-in effects consists in a major deconstruction of the company, basically splitting the current organization into two separate entities (section 4.4.2). Within this new dual organizational structure, one firm will represent the platform, i.e., the data transport, will basically inherit the majority of old telecom business models and will function in the manner of a utility model. The second company will be built around service models, and develop vertical solutions, i.e., the value-added services, independently from the platform business model, but obviously still on top of it. Hence, splitting up the company into this dual approach might spark off a variety of new complementary assets and synergetic capabilities, when the platform part of the corporation will explore the horizontal broadness of convergence, and the service part will exploit the vertical, specialized customer needs.

Being traditionally a rather transport-oriented carrier, the business model of firm δ_8 is not particularly disrupted by vertical attackers, as the firm focuses solely on the platform, and does not compete with vertical services. Although technologies that are disruptive to the company do emerge, the firm sees a long-term capability based on its solid platform. For instance, VoIP was identified as a technology with the potential to cannibalize margins on telephony networks, but on the other hand just representing another form of data from the perspective of δ_8. Although telephony services fundamentally change in provisioning and pricing models by moving away from circuit-switching towards IP-based data streams, the underlying mechanism is still about transporting bits and bytes. Providers of such new VoIP-based solutions are hence still dependent on an underlying platform as a reliable utility, such as the network owned and maintained by δ_8. By focusing on the horizontal broadness, performance and brand of the platform, the firm δ_8 can hence position itself as an integrator to such complementary vertical development, i.e., through leveraging complementary assets when new vertical technologies emerge.

> "These others are not really competitors, rather complements. They are small businesses who develop VoIP etc., you have some specialization going on, and at customer premise. These small firms inter-

> pret applications in such a way, that new businesses begin to see the benefit of combining through a convergent platform. They have to go and find a company like [us] for handling and distributing traffic." (informant 9, case δ_8)

In the case of firm δ_6, the fragmented and diversified technology base of the different business units and customer segments initially posed a major challenge for building the convergence case for the company. In particular, the firm experienced difficulties in identifying any source for a distinctive horizontal capability, that could leverage the platform advantage in the commercial combination of PCs, printers, scanners and digital cameras. As a common denominator for aligning the firm's competencies into a convergent umbrella could not be successfully created internally, the innovation had to "take place outside of the beast" (section 3.4.2). After having identified key requirements for horizontal capability development, the company was able to search and acquire a platform company, which happened to provide this missing horizontal link. The acquired firm's platform bridges several businesses of the firm δ_6 within a spanning platform, creating a business model, which in turn creates a need for the products of firm δ_6. Hence, the firm δ_6 was able to acquire external capabilities, which over night turned the previously rather distinct assets of the company into a highly complementary offering.

Similarly, firm δ_2 internalized platform capabilities for the mobile business segment through the acquisition of firm β_2, which provides a horizontal link between δ_2's existing products and solutions, and additional further opportunities to develop more targeted and specific vertical applications.

> "β_2 brings a nice experienced company, that has sold into the enterprise, to a different buyer, has a customer base, has sales experience in the enterprise, has a platform for us to leverage on the e-mail side—but also perhaps then to say: what are the other applications that we can leverage through that platform, that someone could access on their mobile device, e.g., CRM etc. They have an e-mail solution and a device management solution—all those things are needed to create a solution." (informant 4, case δ_2)

Apart from the acquisition, the mobile handset division of firm δ_2 is currently in the process of exploring emerging verticals of the online gaming marketplace on top of a platform product which is somewhat outside of the firm's traditional business areas—a combined phone and online gaming console.[9]

4.2 Orchestrating ecosystems and reciprocal incentives

As observed in section 3.5.2, in the process of convergence, the phase of deconstructing and disaggregating value chains and formation of horizontal structures does not represent the final state of the evolutionary process. Like in any innovation cycle, the phase is followed by a new impetus towards specialization (Fine, 1998). But this one appears to be different: it functions according to new competitive rules, which are completely different from previously established vertical structures. Since the business landscape is horizontalized and platform-powered, initial shapes of vertical structures cannot be recreated, as firms have to coexist with a variety of horizontal competitors from previously distinct industry segments. These horizontal competitors, on the other hand, do not necessarily only compete with eachother. Since value chains have opened-up both horizontally and vertically, competitors might additionally end up in mutual buyer-supplier relationships, which puts firms into a position of competing and doing business together at the same time.

Coopetition, i.e., the strategic stretch between competing and collaborating at the same time, was identified as a key element for managing through the convergence process among the case firms, which poses challenges on partnering and alliance capabilities of firms. As the alliance activity of a firm generally is embedded into the strategic portfolio and coevolves with the firm's organization and environment (Koza and Lewin, 1998), contingent capabilities have to be developed accordingly. However, apart from situational approaches that match specific coevolutionary configurations, there is one underlying rationale that seems to characterize any strategy for managing partnerships and competitors at the same time. Whereas traditional industry structures caused firms to think linearly, i.e., in value chains or supply chains, converging environments urge firms to think nonlinearly, i.e., understanding the importance of ecosystems (cf. Moore, 1993). In particular, firms have to move away from solely promoting their proprietary product or solution within the marketplace, as the deconstruction of previously vertical structures renders the potential customer

[9] This combination of mobile application and online gaming represents an example for the emerging need for modification of industrial policy and regulation in response to the convergence phenomenon (table 2.3, p. 39).

base highly fragmented and complex. Rather, they have to promote the convergent ecosystem, i.e., invest in assets that are beyond the boundaries of the firm, and create benefits not only for the investing firm, but even for all other firms participating in the ecosystem.[10] Since the value-add of convergence, however, consists of the applicational synergy of previously distinct technology domains, it is highly unlikely for a single player to achieve this transition among customers, suppliers, partners, and other stakeholders. Without investments into an ecosystem—a community of stakeholders which eventually might benefit from the convergence process—there will not be enough incentives for other firms to take part in this synergy. But once the ecosystem has reached its critical mass, and the convergence applications provide superior performance towards end customers, this inevitably creates a demand for firm-internal capabilities, which due to the networked nature encapsulates higher growth potential than any single firm could achieve on its own. In other words, it is about collectively creating an environment where complementary assets fall together.[11]

Hence, when industry boundaries become blurred, partnering and alliance management is manifested in firms' capabilities to manage the ecosystem with its reciprocal incentives. Taking this even further as a rationale for developing managerial action, positions of power or dominance can hardly be based on solely administrating a portfolio of horizontal and vertical partners. The crucial differentiator may be in a firm's capability to *orchestrate* the ecosystem, i.e., the ability to strategically induce actions into the network, which indirectly yield beneficial outcomes for the firm (cf. Gawer and Cusumano, 2002).

In developing their distinctive position within the emerging ecosystem, and in shaping capabilities for orchestration, firms' strategies have to find a proper balance between exploring and exploiting alliances (cf. Lavie and Rosenkopf, 2006), which in the context of convergence can be observed as the need to gradually evolve from the

---

[10] Sticking to traditional and established rules of the industry structure and lapsing into a self-reinforcing direct competition might in fact lead to drastic consequences, which can be described as the "red queen effect" of competition. When an industry is about to change, a strategy of "trying harder" might lead to failure, and it is in such situations important for firms to be alert to respond to the new competitive constellations by fundamentally changing existing business models (Voelpel, Leibold, Tekie, and von Krogh, 2005).

[11] Similar observations have been made in the pharmaceutical industry sector (Erat, 2004). Furthermore, this can be regarded as a special form of private-collective innovation incentives, which takes place at the firm level, and where the emerging ecosystem community shares an open, collective, and distributed business model (cf. von Hippel, 2005; von Hippel and von Krogh, 2003, 2006).

one to the other. In the comparison of the cases, it was observed that, regardless of firm size, during earlier phases of the convergence process, firms have to collaborate and compete in such a way, that allows them to learn and acquire capabilities from each other (i.e., exploring alliances, cf. Lei and Slocum, 1992). In latter phases of the convergence process, the target should be at maximizing utilization of complementary assets, as each party brings and contributes a distinctive capability in a particular fragment of the value creation synergy (i.e. exploiting alliances, cf. Lei and Slocum, 1991). However, firm size and history matter in the way specific complementary assets can be leveraged in such a path-dependent transition from exploration to exploitation, which provide the rationale for a further contingent analysis.

Based on the observations as elaborated in the following sections (4.2.1–4.2.4), a set of capability elements is identified in contingency to the coevolutionary classes. These elements are summarized in table 4.2, and yield the following proposition:

Proposition 10. (P_{10}) *Contingent on its coevolutionary class, a firm needs to develop a specific dynamic capability for understanding, anticipating and orchestrating an ecosystem of coopetition and reciprocal incentives.*

Table 4.2: Capability elements for ecosystem management

Pioneering disruptors	*Vertical attackers*
leverage neutrality advantage	from supplying established firms to competing with them
grow into new rules of competition	induce established firms' dependency on own assets
anticipate emerging ecosystem	disrupt ecosystem, break rules
Platform consolidators	*Reincarnating giants*
leave vertical integration behind	„chinese wall", sharp distinction between own and shared intellectual property
co-shaping new rules of coopetition	position oneself as experienced, reliable partner towards entrant firms in ecosystem
aligning innovative activities into emergence of ecosystem	harness ecosystem for launching new innovation cycles

4.2.1 Pioneering disruptors: coevolve with ecosystem

Firms within the class of pioneering disruptors find themselves in this early phase of exploiting the new rules of coopetition—although it in their case does not come as any major shift. Instead, being founded at this early stage of the convergence process, these young firms rather 'grow up' with constantly being in both collaborative and competitive relationships with larger firms.

As these firms do not carry the burden of tradition, and hence present relatively little path dependency in terms of whom to collaborate with (and whom not), they have a distinctive advantage in terms of neutrality within the ecosystem. There are no real enemies, but not any real friends either. Hence, these firms can build on the neutrality advantage in shaping their role within the emerging ecosystem, and can based on this win significant deals with established firms, allowing the pioneering disruptors not only to gain foothold in the converging industry space, but to strongly influence the economic selection process toward emerging dominant designs.

In this line, a firm like α_5 has the ability to act as a "technology innovator in this way of doing convergence" (informant 17). Without any past foothold in the industry, the firm was recently able to launch a platform product, that "turned into an industry standard with the backing of major handset manufacturers" (informant 17). The informant further reflected, that whereas it normally takes 3–4 years to get a standard, the firm α_5 was able to get it in 18 months, with four major handset manufacturers committing onto that standard. Whereas the firm had the advantage of taking part of the ecosystem right from the beginning, it was on the other hand also a challenge to get this ecosystem into motion. "It's a classic chicken-egg effect, where you get into tricky situations" (informant 17).

The success of the product was dependent on the amount of firms adopting the platform, but on the other hand the first firms were hard to acquire, as incentives to join are low as long as there do not yet exist any other participants in the ecosystem. Hence, the firm had to invest a high amount of resources into educating the market, in parallel to developing the technological prototypes. It was crucial for the firm to keep up a dialog with several major handset providers at the same time, where the firm benefited from the advantage of neutrality.

> "Being independent is really helpful right now. Nobody sees it as a threat. We get to talk directly to the CTOs; they have more willingness to work with us more openly. If the firm δ_2 would do the same thing—you always want to sell what you got—you cannot have an

open conversation about technological directions; and even remote competitors, or carriers, who do not use δ_2 equipment, are not interested. We have hence more ability to synthesize market information and understand market trends. They [i.e., firm δ_2] might have even more people working on the thing. But we understand the market forces better." (informant 17, case α_5)

The firm α_4 finds itself in a similar role within the industry, where neutrality is a prerequisite for gaining foothold in the emerging ecosystem.

"We're neutral. Carriers do not care about that too much. But content providers want neutrality, any content, any device, any network. We have to have interoperability with each carrier because of that." (informant 18, case α_4)

Based on that starting point, the firm α_4 anticipates a future opportunity to disrupt the ecosystem, based on creating "some technique for getting content to the devices, or some technique for building a business model that decreases friction in the value chain" (informant 18). A possible example for this could be seen in delivering advertisement services in a way that is sensitive to the context, thereby establishing horizontal links which leverage synergetic advantages on the one hand, but at the same time changing the rules of previously distinct business domains. However, before being in a situation that would allow the firm to disrupt the ecosystem, they have to focus on interoperability above all—it just "has to work with the ecosystem" (informant 18).

4.2.2 Vertical attackers: rule-making through rule-breaking

The further the process of convergence advances, and hence the new rules of the ecosystem start to gain shape, the easier it becomes for entrant firms to exploit their distinctive relationships with incumbent firms as well as to break the rules of the ecosystem. Firms among the class of vertical attackers seem to find themselves in such positions.

Among them, firm β_5 went into business through delivering the missing link between mobile handset and personal computing applications, allowing end-to-end mobile business solutions. This missing complementary asset did not only facilitate the emerging ecosystem of all involved parties to gain critical mass, the firm was also able to gain such a strong foothold within the new competitive structure, which allowed them to gradually infringe on carriers' business territory. In other words, starting as a supplier of standalone, turnkey solutions to carriers', the firm is today competing with them.

> "In practice, we end up helping the carriers to sell that solution, because it's a little more complex than they're normally used to delivering." (informant 12, case β_5)

Through on the one hand developing the complementary asset for the ecosystem, but on the other hand internally developing distinctive capabilities based on technology, business models, marketing and deployment, the firm was able to induce a strategic dependency from their customers' perspective, yet at the same time moving downstream along the value chain. Based on exploiting the opportunities of coopetition and breaking the rules of the ecosystem, the firm finds itself today in a position where one can get into business with the customers of customers. In the case of firm β_2, coopetition and neutrality have reached a stage where the technological distinction of the platform has made competitors to customers, rendering a relatively unique position within the ecosystem.

> "Actually, I give you another form of coopetition, I guess we would consider [a well-known handset manufacturer] our biggest competitor, and they are actually also one of our biggest customers! So, we have a 120 patents on synchronization technology, nobody is suing us, everybody is suing everybody else. And there is a reason for that, because everybody is not implementing synchronization as a basis for [wireless] push e-mail [...] And the only way you can do that on a [device of the well-known handset manufacturer] today, is by connecting it through a cable and pressing 'sync'." (informant 1, case β_2)

On the contrary, the firm β_1, who enters the market at a very mature stage of the convergence process, does not have any specific ambitions on building-up a specific role within the ecosystem. As a small player, it is practically impossible to create your own ecosystem at this stage, and although standards can be created, it is unlikely to achieve any significant adoption within a short period of time. The company has to "live aside someone else's ecosystem" (informant 15).

4.2.3 Platform consolidators: internalize coopetition

In the class of platform consolidators, the early phases of the convergence process on the one hand, as well as the the size, experience and resources of the established firms on the other hand, allows managerial action to get actively involved into co-shaping the emerging new rules of competition. By leaving established structures of vertical integration behind, and aligning existing proprietary innovation activities into the openness of the ecosystem, these firms can create opportunities for exploiting investments into industry-wide horizontalization. In particular, such opportunities represent the result of actively welcoming the open ecosystem as well as of investing innovation activities into advancing the process of convergence. Based on that, in the longer run, firms may induce a need for complementary assets, which in turn are anchored into their own platforms.

For instance, the firm γ_2 decided to adapt its strategy for being better positioned in the emerging process of structural change. The strategy inflection point was based on the top management coming to the realization, that if γ_2 is going to be a platform company, they will need to welcome partners, customers, developers and other stakeholders to participate in the development of this platform. In particular, although being in on the business-to-business side within the software industry, the firm decided to open-up toward the "megatrend of open innovation, since the Internet is increasingly about participating online" (informant 23). Hence, previously established rules do not hold anymore, as people are not just passive consumers anymore, but authors of content, and reviewers of other people's content (cf. von Krogh and von Hippel, 2006). Inspired by this trend, where customers may suddenly become suppliers and vice versa, the firm γ_2 created an open developers' network, aiming at adopting the new mechanisms of the emerging ecosystem into the firm's own innovation activities. What previously was created, implemented, solved and patched by in-house engineers, is now being allocated into an open network, where the firm's own software engineers have practically the same status as any interacting user within the worldwide network.

Today, the issue of understanding and managing the ecosystem is on top of the agenda at firm γ_2, and the managerial commitments are accordingly high. As part of the initiative, the firm appointed an executive vice president of ecosystems, heading all ecosystem-related strategic issues within the corporation. Hence, through facing the trend at rather early stages of the convergence process, the firm wants to significantly contribute to the success of the converging ecosystem, but on the other hand develop internal capabilities to be ahead of its competitors in capitalizing on the investments into the model of openness.

"So, the idea behind [the developers' network] is to put the mechanisms in place, and then help them along, for the community to help itself." (informant 23, case γ_2)

And by helping itself, the community does indirectly participate in the innovation trajectory of γ_2, since the horizontal platform is being constantly reviewed, corrected, improved as well as extended by thousands of vertical applications on top of the platform—around the clock. Hence, the firm is able to capitalize on complementary assets both horizontally and vertically, with an incentive system based on solely reputational capital.

"But for the most part, people don't win anything. They win a t-shirt. They win glory. They win a, you know, star next to their name. That's about it. And it's amazing to see how motivated people are by the recognition and appreciation of their peers." (informant 23, case γ_2)

Furthermore, the firm managed to capitalize on its strong firm brand in building the reputational system within the developers' network. Since the firm γ_2 itself only participates in the network, but does not attempt to dominate it, it gained credibility within a short time and was well accepted by the first customers, as well as customers' customers. Interestingly, the reputation status as determined through the participation in the network has become an expertise and performance measure even beyond the community, and does even appear inside single organizations. A customer of the firm γ_2, one of the largest systems integrator companies in India, measures itself by being the top contributor in the developers' network based on the total number of points awarded. But even further, the systems integrator company defines employee goals for the year as a certain number of points to be achieved within the network, and conducts bonus payments based on such points. In other words, that customer wants to be known within the ecosystem for having expertise about the platform of γ_2.

Whereas the firm γ_1 invests into the emerging ecosystem as well, it does so rather on firm-level than in terms of integrating end-users into the innovation system. Being "a platform company" (informant 24) the growth and direction of the ecosystem is of high interest for the company, and γ_1 has to get a great deal of partners involved into the ecosystem, in order to get the platform-oriented business model running. Technically speaking, the company provides the hardware as "a standard high volume platform" (informant 24), and then relies on the ecosystem in terms of software built on top of that. Unlike the γ_2, the firm γ_1 does not provide an open toolbox for allowing practically anybody to contribute to the development of the platform. Instead, the firm is active in a variety of technical standardization bodies, aiming at gaining a stake of the ecosystem in a similar way. Also, the firm controls a $100 million fund, which is solely reserved for "ecosystem funding". Based on this budget, the firm continually invests into start-up firms in various stages, with an involvement of specific business units in the funding procedure, in order to detect complementary assets at early stages.

Having been within a horizontal business model for quite a while, the firm γ_4 does not perceive the issue of coopetition as a major recent shift.

> "We're used to it. It has been a fact of life for ever with γ_4. [...] It has been in the nature of the network business for a while. For companies who are vertically integrated, this is new for them." (informant 19, case γ_4)

The firm γ_4 has been involved in a variety of collaborative projects with competitors, and has hence been able to build up a credibility based on that. However, the firm does induce ecosystem effects as well, which for instance was seen through the acquisition of a competitor firm within the cable business. Even after the acquisition—instead of joining forces—γ_4 is interestingly still competing against the acquired company.

4.2.4 Reincarnating giants: cold-hearted commercialization

In the class of reincarnating giants, firms find themselves in rather commercial stages of the convergence process, where the new rules of the ecosystem have nearly been shaped, and the firms have to overcome different kinds of inertial structures for reacting against attacks from entrant firms. As the horizontalization trend reaches rather mature stages, where these firms have to stayed focused on capitalizing synergetic advantages instead of loosing market-share, it can be regarded as rather unfavorable to concentrate on vertical efforts in parallel. In particular, as these firms have a tendency to translate external impulses into strategic inflections rather slowly due to the size, legacy and inertial strategy vectors, it is rather difficult for such firms to be part of—and actively follow—the dynamics of both a collaborative and competitive ecosystem. Provoking coopetition based on suddenly opening-up broad areas of the company's innovation activities might overly expose themselves to strategically vulnerable situations, which might become difficult to navigate.

Instead, when entering collaborations with competitors, a distinction of strategic approaches, depending on size and radicalness of partner firms seems feasible. In situations where an established firm choses to grow horizontally and further develop its platform strategy, collaboration with firms in similar positions may represent an appropriate approach. Typically, this partner would be an established firm itself, hence a competitor, allowing potential to join forces of specific projects, thereby extending both firm's stake in the horizontal structure of value creation. For instance, the firm δ_2 has been involved in various horizontal collaboration projects with competitors, such as firm γ_4, where firms collaborate in bringing a specific technology into the ecosystem, but on the other hand compete within other business areas. As elaborated above (section 3.2), in many cases, these situations may be the effect of the evolutionary development of trajectories, where knowledge bases come together, without involved firms necessarily searching for direct competition.

> "All of a sudden, former companies that were never considered potentially competitors, are cropping up as the [...] biggest potential competitors. Also some due to the fact that some of those competitors, e.g., γ_4, need to be very close partners to us. So, there is a very fine line, and always becoming more blurred." (informant 4, case δ_2)

In such situations, it is crucial to develop a sharp distinction between proprietary, protected intellectual property, as well as shared intellectual property, which may be subject for external collaboration.

> "It's a very delicate situation, you have to set up a "Chinese wall" between the two so that they are not sharing information, because you have to protect that intellectual property against the firm within, say within γ_4, that you're working with. They are not going to compete against you, but at the other business, they have got a technology, they may develop a solution that competes directly with you." (informant 4, case δ_2)

On the contrary, setting up this "Chinese wall" may become rather troublesome when dealing with a variety entrant firms, as they may start to compete with the established firm in an unexpected and unimaginable manner, where a priori protection of intellectual property might not guarantee any strategic protection. Instead, in order to leverage the innovativeness and radicalness of attacking entrant firms, without compromising established and profitable business models, established firms may find ways to transform their incumbency into a complementary asset, instead of a rigidity. Aspects such as market power, experience, brand and reputation of the firm may be translated into a distinctive capability in such a way, that it makes the established firm an attractive and beneficial partner to a variety of entrant firms in the ecosystem.

Staying with the case of δ_2, the firm is currently undergoing a slight strategic reorientation towards more service provisioning-oriented, i.e., vertical, business activities, in response to the commercial stage of the convergence process. The firm has among others been trying to offer solutions within the mobile e-mail service segment, which historically has been covered by smaller firms, such as firm β_2. Hence, the firm δ_2 and the set of entrant firms have until recently been regarded as competing within the same market, despite the different magnitudes and underlying business models of the firms. Many entrant firms have actively been establishing contacts to firm δ_2, as they were aware about δ_2 desperately trying to develop a proprietary standard with minor success. Based on this demand of collaboration from the side of the entrant firms, as well as due to the resulting competitive pressure, the firm δ_2 decided to turn its structural inertia into a favorable complementary asset. It recently launched an aggressive partnering strategy, deciding to address the customer demand in this service segment through collaboration with primarily eight small and medium-sized enterprises within an underlying larger strategic network, or forum. Since the service segment is fragmented into many sub-niches, all of which target specific customer demands, the established firm was with the help of these partner firms able to offer a broader solution portfolio through orchestrating its partners instead of attempting to manage all technologies, thereby avoiding integration or internally building all required competencies. At the same time, the small players perceived themselves to be at the right place when δ_2 made a commitment on actively participating in a value network, which in the case of β_2 later even lead to an acquisition.[12] Interestingly, the firm δ_2 gave up the plan to develop a de facto standard in-house for this service, a plan that originally was envisioned to gain a dominant design based on the existing market share, which further could have shaken out several small solution providers from the market. But alone, the δ_2 was simply not able to leverage the syn-

[12] From the perspective of β_2, this can be additionally explained by the strategic focus of the firm, committing onto partnership management as a key source for competitive advantage. The firm has been creating opportunities through intelligent partnerships, constantly re-assessing and reconfiguring the value network, instead of leaving it on a static basis.

ergetic effects of convergence, and decided to become world-market-leader by offering a broad variety of various specific solutions under its umbrella brand. And as the convergence trend implies a need for vertical solutions,[13] yet at the same time making it impossible for incumbent firms to regain vertical structures, the firm managed to adapt to new forms of vertical consolidation, i.e., through coopetition. Based on this partnering strategy, both δ_2 as well as a set of involved entrant firms were not only able to develop industry foresight in terms of understanding future market needs in more detail, but also to actively co-develop radical solutions at the intersection of technologies. The networked innovation approach created a win-win situation for both companies, allowing co-creation of value in a converging environment, through access to a broad variety of industry actors on the one hand, and the access to a wide portfolio of customer demand on the other.

Similarly, instead of entirely exposing oneself to the unpredictability of the ecosystem, firms may go for a venturing approach, with the aim of rather experimenting than entirely adopting the rules of coopetition. In other words, firms might rather attempt to exploit opportunities within the dynamics of an ecosystem by incubating spin-off firms, which would go out and act on behalf of its mother company within such an ecosystem.[14] For instance, through the acquisition of the platform company, which serves as a horizontal convergent integrator between the firm's different business units, the firm δ_6 does not only exploit internal complementary assets, but can even induce exogenous effects within the ecosystem in an almost disguised role of a start-up firm. In parallel to running own vertical business models on top of that platform, the firm licenses the same platform to competing third-party providers of such vertical applications.

> "In that sense, we are creating competitors. What we do on the online side as well, is, we create competitors from [our platform's] perspective, we power [a variety of retail stores, who develop vertical applications based on our solution]. They all have their separate online site, that is separate from [our] users. So, we have licensed to those

[13] In the case of mobile e-mail, customer demands have become increasingly specific and complex, requiring rather tailor-made, end-to-end solutions, than standard platforms.

[14] A similar behavior, i.e., getting "a man inside" in order to gain access into open communities where one is not present, has been observed on a user-level, where some firms sponsor people to act strategically in open communities, aiming at secretively exploiting freely available complementary assets (cf. Dahlander and Wallin, 2006).

people this platform so that they can compete. And they get all the innovations, that [our platform] gets, and they simply need to rely on their customer relationships to acquire users. And when someone acquires [such a] user, typically, they are doing so at the cost of [us] acquiring that same user." (informant 16, case δ_6)

Hence, the firm δ_6 fosters competition on the one hand, but on the other hand, acting as an entrant firm into customers' established businesses, the firm is able to break the rules and exploit the new dynamics of the ecosystem.

Returning to the case of firm δ_2, within another business segment, the approach towards platformization did not go through acquisition of a start-up firm, but rather through the collaborative foundation of an industry consortia together with other incumbent manufacturers. Whereas the achieved platform standard in technological functionality and in its diffusion among several vendors creates interoperability that speaks for itself, certain observations of exploiting the open innovation system based on a dominant role in the industry could be made from the perspective smaller firms. Formerly having worked with the consortial platform organization, the informant 15 explained:

"[The consortium] in theory, was created by the idea, that a bunch of manufacturers would come together, and create a platform, that would lower the cost, of manufacturing for all, and would provide a single platform for applications. [...] But it became clear early on, that was the reason why I left, [...] and many others left, that the firm δ_2 just wanted a shared engineering organization, where they dictated the specifications." (informant 15, case β_1)

A hybrid approach between collaborating horizontally and opening-up towards vertical entrants can be observed in the case of firm δ_5. For the last several years, the company has been in a variety of horizontal partnerships allowing the firm to sell more devices and device components (e.g., webcams). But the firm has also taken some risks, when it partnered with a large online media provider, which from the perspective of δ_5 can be seen as a vertical partner, and which based on the structure of involved distribution channels indirectly made the firm's technology available to all competitors. At that time, the firm decided to do so in order to extend the market and the ecosystem, with the risk of vertical firms infringing on established margins. After all, the firm pursues a strategy of retaining a solid horizontal platform, which will be massively scaled as new vertical business models emerge. In order to achieve this growth advantage, the firm does not hesitate to invest into the ecosystem and induce actions in favor of competitors, as long as the ecosystem, and thereby the demand for the platform, keeps on growing.

> "I think that the beauty of creating a platform is that the ecosystem will shape up itself the way it wants based on the innovation of each of the companies around them. [...] As a technology provider, you're very likely to try to partner with the leader. And then, there is a sort of self-fulfilling prophecy, the more you become a leader, the more people want to gravitate around you, because the more success they are likely to have as well. Success brings success." (informant 7, case δ_5)

4.3 Protecting complementarity against redundancy

As seen in the previous section, depending on the role within an ecosystem of convergence, required capabilities for managing collaboration, competition and orchestration may have highly different shapes. Based on the positioning within the coevolutionary classes, firms can exploit the specific constellation for building specific distinctive strategy vectors, which make them inimitable to competitors outside that respective class. As the process of convergence erodes boundaries between previously distinct knowledge bases, firms may search for roles of either producing or inducing a channel for leveraging complementary assets between these bases, which provides a path for sustaining the existence and growth of the company.

As the new rules of the ecosystem have become highly complex and even more difficult to predict, initiatives for dealing with these new mechanisms have to be linked with highly flexible development paths, allowing room for strategic changes in response to the coevolution. In particular, firms have to manage the ambivalence of finding a specific—either horizontal or vertical—niche for exploiting complementary assets within, and at the same time avoiding coevolutionary lock-in or lock-out, which eventually might lead to obsolescence of the entire company. As the convergence process brings formerly distant industries closer, complementarities might well turn into substitutes. Hence, any convergent strategy should be characterized by both 'a way in and out', i.e., building strategy vectors that leave room for path flexibility instead of path dependency.

In the case of entrant firms, a general strategic approach for exploring and exploiting the process of convergence can be observed in the role of acting as a mediator between the emerging gaps of convergence. These gaps generally occur through the serendipitous and coevolutionary spill-over across industry boundaries (section 3.2.1), where potential for combining elements of previously distinct technologies and business models arises. Such mediating business models operate horizontally, vertically or even both, and link customers or businesses who are interdependent, or would like to be. Furthermore, these models are in most cases based on the advent of underlying mediating technologies, which enable the implementation of the business model on a technological level (cf. Afuah and Tucci, 2003). Hence, the application of mediating technologies can serve as an elementary mechanism for capitalizing on intermediate stages of the convergence process. Typically, platforms can be practically realized through mediating technologies, such as for instance middleware, adaptors, translators, and other application models that become technologically possible due to the modularity of the industry (cf. Baldwin and Clark, 1997). Hence, based on mediating technologies, firms can develop corresponding business models, which connect at two or more previously distinct points within the intersection of industries, thereby capitalizing the potential of spill-over effects into value creation.

However, the role of mediation has a limited life time. As long as the convergence process is advancing in its early stages, where vertical structures become deconstructed and new opportunities for technological recombination emerge, finding and building a mediating business can be a highly profitable approach. On the other hand, when the convergence process starts to mature, the industy-wide spill-over effects become increasingly translated into business models in a variety of—both entrant and established—firms, who have invested resources into developing and acquiring competencies from new knowledge bases. For mediating firms, the pressure by the ecosystem is likely to increase, as established 'follower' firms attempt to regain vertical structures and induce actions through either partnerships or acquisitions, which render the mediating business model obsolete.

For instance, consider a mediating firm which bases its business model on interconnecting two established firms from previously distinct industries. This may represent a highly viable solution as long as the established firms remain in different worlds, since under such circumstances, it is solely the mediating firm who creates and captures the value of combination, as it delivers the synergy of both worlds. However, through its pioneering role of creating value at new places within the ecosystem, the firm is actively driving the trend of value chain deconstruction and vertical disintegration, thereby inducing a variety of competitors—including the established firms themselves—

to respond to this trend. Even though the entrant firm may have an early-mover advantage, the established firms may attempt to infringe on their business model. This may happen either through diversification or specialization into technologies that provide the mediating adaptor in-house. But also, the two established firms may sign direct mutual partnership deals, which lock the mediating firm out, or may even merge into a single company, acknowledging that the convergence of the industries can be optimally exploited through entirely merging the originating businesses. Hence, if the established firms suddenly may deliver the synergetic value of both worlds by themselves, the sudden erosion of competitive and corporate boundaries may cause a lock-out to the mediating firm, as its business model has become irreversibly redundant. Although mediating business models and technologies represent a highly viable way of developing complementary assets being an entrant firm, they are likely to experience creative destruction (Afuah and Tucci, 2003).

Accordingly, in order not to get "stuck in the middle",[15] yet at the same time being able to drive and capitalize the spill-over between industries, entrant firms have to develop capabilities of firstly identifying an emerging gap of missing complementary assets between established business models, which serve as the foundation for creating a mediator role. Secondly, meanwhile the entire firm may be based on delivering the elements of such complementary assets, e.g., mediating technologies, the firm has to develop, maintain and internally communicate a clear understanding of the potential life-cycle of that value proposition. As the value proposition may be unstable, yet profitable, these firms have to create an awareness of potential external factors that may suddenly render their model obsolete. In particular, mediating firms have to permanently balance between the ability to sustain the distinctive complementary assets throughout the industry development, as well as the ability to acknowledge their end and to enter an alternative, post-convergence path. In some cases of mediating business models, a long-term viability of the complementary asset may be possible through creating established firms' dependency of the mediating elements. For instance, small firms may increase the inimitability through technological complexity, combined with a strong relationship to end-customers, which makes it difficult for established firms to abandon the mediating firm's participation. Hence, mediating firms

[15] The strategic positioning of being "stuck in the middle" was introduced by Porter (1985, p. 12), and can be characterized by below average performance and low profitability.

may be able to harness complementarity on the one hand, but at the same time inducing a lock-in from the perspective of established firms.

In other situations, however, this lock-in of established firms may be rather hard to achieve, especially if there are mediating competitors and the technological capabilities *per se* may be less distinctive. In such cases, alternative paths have to be predefined and ready to execute. Such paths may either consist of incremental strategic development, such as technology and product roadmaps, which aim at keeping the mediating firm one step ahead of the established firms as well as growing the company into new generations of the mediating role. On the other hand, such a plan may leave room for major strategic inflections, where either entirely new areas of mediating business models are detected or the firm maintains a proper way out of business.

Hence, based on the observations as elaborated in the following sections (4.3.1–4.3.4), a set of capability elements is identified in contingency to the coevolutionary classes. These elements are summarized in table 4.3, and yield the following proposition:

Proposition 11. (P_{11}) *Contingent on its coevolutionary class, a firm needs to develop a specific dynamic capability for exploring and exploiting complementary assets that leverage the inter-industry spill-over, as well as for anticipating and reacting on the limited lifetime of such assets.*

4.3.1 Pioneering disruptors: mediate and differentiate

In the class of pioneering disruptors, a general tendency of mediating and specializing could be observed. At this explorative stage of the convergence process, the core strategic focus generally centered around exploring the market demand based on the gap between two worlds and building the business case in-between. Firms such as α_6, α_5 or α_4 are currently in the process of identifying and testing the business case for the respective complementary products, which make use of existing conflicts between the previously distinct industries. In addition to that, the firms α_3, α_2 and α_1 are currently developing technological capabilities, which mainly are reflected in the respective highly-complex and multifaceted platforms. Also, through establishing a wide and heterogeneous set of external relationships—i.e., customers, suppliers and competitors—these firms anticipate the emerging pressure by established firms, and work towards the inimitability and future stand-alone viability of the mediating model.

Table 4.3: Capability elements for sustaining complementarity

Pioneering disruptors	*Vertical attackers*
find complementarity niche	protect niche advantage; avoid becoming obsolete
mediate between established business models	post-convergence path; further distinguish complementarity or exit
anticipate emerging conflicts between established business models	induce dependency and lock-in of established firms towards mediating model
focus on inimitability of assets	
establish heterogeneous set of external relationships	
Platform consolidators	*Reincarnating giants*
find complementarity dominance	protect dominance advantage; avoid becoming commoditized
develop scale advantage; make too costly to imitate	build margins on (new) combination of assets not on assets *per se*
value network assimilation	value network orchestration; induce incentive and lock-in of entrant firms to participate in own innovation

4.3.2 Vertical attackers: induce dependency or exit

As the convergence process proceeds into commercial stages, the pressure by established firms onto mediating firms increases, as these attempt to imitate or launch alternative business models, in order to compete with the smaller firms. In such situations, it is crucial for mediating firms to find ways to protect the niche advantage, and avoid becoming obsolete through creative destruction. Firms such as β_5, β_4 and β_3 have apart from the technological expertise been strongly involved in shaping respective emerging ecosystems, where established firms see stronger incentives for participating according to the new rules of the game, instead of imposing their respective dominances through attempting at regaining vertical integration. Apart from that, as indicated though the case of firm β_6, entrant firms may benefit from lower structural inertia and internal reluctance of change, when introducing changes to the business model. On the contrary, the firm β_2 regards its current mediating business model as exploited, but leverages the underlying mechanisms of the technological platform for a new generation of applications, which provides the company with a growth path into emerging techological cycles.

4.3.3 Platform consolidators: complement but orchestrate

Whereas entrant firms may generate viable strategic approaches in the concept of mediating technologies and business models, established firms have to play the same rules, but from the opposite side. As the process of convergence erodes boundaries between previously distinct industries, established firms need to find ways of securing their industry dominance and independence among the emerging set of entrants, i.e., not being completely *mediated*. Whereas a beneficial strategy for entrant firms is based on identifying complementary niche areas, i.e., complementing internal assets with external ones for rendering a novel solution, established firms may look for it inside. In other words, this type of convergent capability is in most cases based on the identification of synergy potentials within the company, and the ability to leverage complementary assets across divisions of the firm, allowing the firm to grow horizontally with the convergence process. On the other hand, as the external environment coevolves, and eventually external vertical business models emerge on top of horizontal platforms of the established firms, there is a risk of redundancy for these firms as well. Although the entire companies may not drift into obsolescence in the same manner as entrant firms might do, there can be tendencies of strong commoditization, as entrant firms use the established firms' horizontal structures as a vehicle for their own vertical initiatives. In such situations, the inherited structures and the size of the company may inhibit established firms from cannibalizing their own business through introducing similar vertical business models, or changing direction of the business.

Hence, in early phases of the convergence process, established firms may not only base their capability on leveraging complementary assets within the company, but also on identifying methods to reflect their previous dominance in the new model. Like in the cases of firms γ_4, γ_1 or γ_2, the previously fragmented internal knowledge base and technological portfolio can be rearranged allowing cross-fertilization and horizontal broadness. Additionally, due to the market power and dominance of these firms, the integrated platform approach can be easily scaled, and thereby becomes too costly for smaller players to imitate. Also, the install base at customer premises can serve as a vehicle for reducing lock-in effects in such situations. Additionally, these firms need to develop a sense on how the emerging ecosystem affects their future roles, thereby accepting to regard elements of the convergent platform as a part of the value network.

4.3.4 Reincarnating giants: protect incumbency

Investments in value network activities become even more important, as the convergence process advances into mature stages. As seen in the cases of δ_7, δ_6 or δ_2, the initial strong foothold within the industry can be transformed into a position of being a strong value networking partner, allowing the established firm to orchestrate smaller partners. Based on such an approach, these firms gained capabilities to induce an incentive and a corresponding lock-in of entrant firms to participate in the innovation agenda of the established firms themselves.

Also, the horizontal axis of collaboration, i.e., partnering activities with competitors, represents a sensitive aspect from the complementarity perspective. Whereas opening-up parts of the company on the one hand promises opportunities for further developing towards the horizontal broadness of convergence, other parts of the company may be directly competing. In such situations, possible dominance advantages in one business segment have to be protected from spillovers from other segments—either through entrant or other established firms—who may commoditize parts of the company. As discussed above (section 4.2.4), the firm δ_2 has implemented a strategic initiative for orchestration of decentralized innovation structures.

> "At δ_2 , our approach is to think: what do we bring to the table, what intellectual from hardware and software side, what solutions do we bring to the table that create intellectual property for us that will be important for other partners, [...] They bring something to the table, once we sell into a marketplace, and into an enterprise, e.g., we have created a spot for ourselves, where people can compete against us, but they can't come in and make us a commodity." (informant 4, case δ_2)

On the contrary, in the cases of carrier firms such as δ_9, or δ_1, which represent large and traditional organizations with broadly inertial technological infrastructures, firm-internal complementary assets may be difficult to exploit in a reasonable time. As the value creation has moved downstream, their current business models have increasingly become commoditized into solely data transport (section 3.5.3). During earlier phases of the convergence process, these firms did not search for complementary assets or opportunities in collective action, but did instead rely on their respective dominant positions in the underlying distinct industry segments. But as the value creation potential shifted towards the combination of assets, away from the assets *per se*, these firms find themselves increasingly locked into their existing roles, and in particular, locked out of downstream value generation opportunities.

Finally, as seen in the case of firm δ_5, mediating business models must not necessarily be limited in life time, but can serve as a model for sustaining even longer-term growth. However, in that case, the mediation does not happen between multiple technologies or business models. Technically, as the firm's value proposition is to provide the man-machine interface for a variety of current and emerging applications in the realm of convergence, the company is dealing with an almost timeless concept of mediation technology.

> "As long as there is human, who has to interact with the machine, we look at how can we actually put ourselves in the middle." (informant 7, case δ_5)

4.4 Reorganizing the firm for convergence

Particularly large firms, which can be characterized by the corporate legacies inherited from respective isolated industry fields, have to overcome various facets of inertia for leaving the burden of tradition behind. When responding to new rules of the market, as implied by the drastic shift of competitive boundaries due to industrial convergence, the existing internal structures, consisting of competencies, processes, technologies, etc., may not be able to fulfill the new requirements on the firm. Besides exogenous determinants of this change, such as technological trajectories or the coevolution of firms, path dependency and inertia are often locked inside existing organizational structures, that inhibit the firm from moving into radically new directions. In particular, the firm's responsiveness to the potential of emerging exogenous spill-overs between knowledge bases may be limited by the way a company is organized internally. Among the set of established firms analyzed within this study, the observations yield that reorganizations are necessary in many cases, where capabilities for bridging previously distinct competence islands into synergetic, but also unpredictable, effects represent the prospect of managerial search.[16]

---

[16] Any particular observations regarding organizational changes within the set of entrant firms could not be made. Recognizing the fact that such, mostly young firms do not possess a history long enough for allowing any major reorganizations, the class of pioneering disruptors and vertical attackers will not be further elaborated within this section.

As boundaries between firms' internal and external activities blur, and new forms of collaborating and competing occur, the emerging structures of the overall ecosystem have potential to give rise to promising new modes of organization for innovation (cf. von Hippel and von Krogh, 2003). After all, organizational structures can be characterized by their intrinsic need to continuously adapt to their respective environment (Lawrence and Lorsch, 1967). Organizations seek to become isomorphic—similar in form and relations—with their contexts (DiMaggio and Powell, 1983). Hence, the notion of *organizational convergence* is in this work used to refer to not only rearranging and merging internal structures, but to introducing even higher level organizational constructs and agendas, which aim at addressing previously uncharted forms of internal spill-over, and thereby creating new capabilities for innovation.[17]

As a consequence of convergence, firms need to be able to adapt and extend organizational structures in such a way, that the linkage between internal knowledge bases of the firm is able to dynamically anticipate, follow and match the new structures of the industry. In undertaking such organizational change in accordance with environmental impulses, one can generally distinguish between two levels of adaptation (Argyris and Schön, 1978; Bartunek, 1984; Tushman and Romanelli, 1985; Watzlawick, Weakland, and Fisch, 1974):

Definition 16. (First-order adaptation) *Being viewed as a local form of adaptation, the concept of first-order change refers to incremental modifications in present ways of interpretation.*

In such a degree of change, firms have to work out specific choices within a given organizational form (e.g., policy changes, launching or withdrawing new development programs, creating new processes). This is in contrast to the following form of adaptation:

Definition 17. (Second-order adaptation) *Second-order adaptation refers to radical and discontinuous shifts in interpretive schemes, which evoke change in the underlying structure itself.*

In other words, this latter type of change does not happen within an organization form, but fundamentally changes it (e.g., chang-

[17] In literature, the term "organizational convergence" appears in a variety of meanings. However, when comparing different facets of this generic construct, there appears to be a rather thin line between the isomorphic notion, i.e., converging *with* the environment, as well as the aspect of internal adaptation, i.e., converging *for* the environment (cf. Bauer *et al.*, 2003; Greenwood and Hinings, 1996; Lant and Mezias, 1992; Lind and Zmud, 1991; Tushman and Romanelli, 1985; van de Ven and Poole, 1989, 1995).

ing from unitary to matrix-organization, creating new organizational structures that substitute previous ones).

Based on studying the set of established firms along different stages of the convergence process, with their specific urgencies to overcome structural inertia, general observations about the development of organizational capabilities have been made. It is argued, that although an immediate adaptation of second-order would be the most suitable for responding to the emerging radical spill-over effects in any situation, it is however not entirely necessary, as long as the firm finds itself in earlier stages of convergence. On the contrary, firms need to find a proper balance between first-order and second-order adaptation as long as the process is in the explorative phases. As soon as commercial phases of the convergence process come into play, and a convergent dominant design starts to gain shape, the managerial commitment will have to shift towards solely second-order adaptation.

Based on the observations as elaborated in the following sections (4.4.1 and 4.4.2), a set of capability elements is identified in contingency to the coevolutionary classes. These elements are summarized in table 4.4, and yield the following proposition:

Proposition 12. (P_{12}) *Contingent on its stage within the convergence process, an established firm needs to develop a specific dynamic capability for internalizing structural industry changes into its organization through first-order or second-order adaptation.*

Table 4.4: Capability elements for organizational adaptation

Platform consolidators	*Reincarnating giants*
first-order and second-order adaptation	second-order adaptation
explore ways of creating organizational convergence; try new organizational combinations	exploit structures of organizational convergence; rebuild organizational entities based on industry evolution
technology-driven adaptation of organizational processes	market-driven adaptation of organizational processes
identification of internal R&D synergy potential	identification of internal and external commercial synergy potential

4.4.1 Platform consolidators: organizational exploration

As long as a firm finds itself in earlier phases of the convergence process, the technological uncertainty is still rather high (cf. Tushman and Rosenkopf, 1992). In such situations, it is crucial for firms to develop distinctive capabilities based on the advantage of timing, i.e., exploring novel ways of internal organizational combinations, and gradually changing the organization along the trajectory of convergence—without taking too much risk at once. This is due to the fact that the emerging environment, for eventually adapting to, does not exist yet. Hence, it is rather feasible to find a proper balance between implementing first-order adaptation and anticipating future requirements for second-order adaptation, instead of already fully committing onto the latter one. Also, as the market is not quite there yet, the needed adaption of organizational processes seems to be rather technology-driven, aiming at identifying internal R&D synergy potential.

Like in the case of firm γ_2, the commitment of changing existing structures from product sales towards ecosystem promotion, may have certain second-order tendencies, but is mainly based on adaptation within the existing organizational framework. The company wants to face the emerging trend of open innovation and reciprocal incentive systems, but as these rules are still emerging, the developers' network initiative coexists in parallel with the remainder of the organization, instead of radically changing or substituting it at this stage.

Similarly, in the firm γ_4, "pockets of innovation" (informant 19) within the organization are continuously screened and identified, and new organizational structures are created primarily on an 'ad hoc'-basis. Hence, in parallel to the static organizational structure, the firm is technically run based on collaborative, multifunctional and multidisciplinary teams, which are clearly communicated and incentivized within the company.

On the contrary, in the case of firm γ_1, the balance towards second-order adaptation can be observed as somewhat higher. Although the firm initially responded to the lack of internal spill-overs through supporting several cross-unitary and cross-disciplinary projects, this development has lead somewhat further into the organizational foundations of the firm. Based on the firm's vision of being a "platform company" (informant 25), the horizontalization has taken place on organizational structures as well. Although the company previously was organized in specific strategic business units, all with their own respective sales and support staff, the company has gradually shifted towards a matrix organization. Whereas one dimension is represented by technology and product organizations, representing specific silos of expertise, the second dimension, consisting of the platform organizations, represents the horizontal linkage for creating synergy and allowing the basis for convergence-based solutions.

4.4.2 Reincarnating giants: organizational exploitation

In more mature, and thereby commercializing phases of the convergence process, established firms are faced by an increased urgency to internally coevolve according to their environments. As the established, inherited metrics based on which the firm operates tend to become outdated as new mechanisms of industrial verticalization evolve, a resulting need for fundamental internal change becomes likely. Hence, serious organizational adaptation is needed in order to navigate the firm out of any potential coevolutionary lock-in. In other words, firms in such situations may find a beneficial starting point in directly facing the urgent need for second-order adaptation, aiming at gaining a more innovative organization, instead of deliberately forcing renewal within the organization. In such situations, it is furthermore mostly inevitable to obliterate any efforts of already initiated first-order adaptation (Tushman and Romanelli, 1985), which on the one hand results in squandered resources, but on the other hand paves the way for large-scale changes, where smaller initiatives may cause unwanted friction.[18]

Hence, instead of solely experimenting with novel organizational constructs, the maturity of the convergence process will imply a need for established firms to partly rebuild organizational structures, in order to translate the external spill-over effects into internally beneficial combinations. Also, as the emerging ecosystem starts to gain shape, and increasing industry-wide mechanisms towards vertical structures will occur, firms will need to seek structural similarities—i.e., isomorphisms—between the organizational structure and the demand from the market.

In doing so, the firm δ_9 has previously initiated a variety of distributed, multidisciplinary and shared projects throughout the entire organizational structure. These have mostly been associated with the

[18] Taking a strictly evolutionary perspective, it can be argued that selection tends to favor higher-performing organizations. Hence, if second-order adaptation generates any gains in performance, it can be assumed that organizations that do engage in second-order adaptation will have an evolutionary edge over organizations that do not (Ethiraj and Levinthal, 2004).

existing matrix-based organization of the company, which bridge different business units of the company with specific projects. The problem, however, that the company still has to overcome, is based on various conflicting incentives, which relate to issues such as project ownership, margin participation, as well as projects which as an outcome might render entire parts of the company obsolete (section 3.4.1). In other words, the firm perceives the existing set of first-order adaptation initiatives as not necessarily leading the company towards more convergence-oriented output. After all, being an incumbent telecommunication carrier, and thereby specifically affected by the disruptive effects of industrial convergence, the firm will need to face more fundamental changes within the roots of established structures, in order to achieve a certain reincarnation of the entire company. Hence, the firm's management is currently assessing ways of second-order adaptation in response to the emerging platform paradigm of the industry. As described above (section 4.1), the plan is to split the company into two separate organizations—utilities and services. In such a changed organizational framework, the utilities organization would represent the platform company, spanning the convergence space and creating horizontal synergy. Based on this foundation, the services organization would represent a vehicle for launching customer-specific commercial initiatives, which represent the vertical applications on top of the platform. In other words, the company is currently in the process of building capabilities based on increasing the degree of isomorphism between the organizational structure of the company as well as the newly shaped convergent industry structure.[19]

Also, in this search for isomorphism through second-order change, the firm δ_2 made a large effort for reducing redundancy between strategic business units. As initially the underlying knowledge bases, and at this point even industries have come together, the firm initiated a deeply rooted reorganization, with the intention to, as described in a

---

[19] These observations complement earlier contributions in literature which argue that platform success needs organizational adaptation (cf. Dürmüller, 2006; Gawer, 2000). Further motivation for the described isomorphic adaptation can be found in the work by Cusumano (2003a,b), who argues that firms need to balance between being product organizations, striving towards leading or complementing platforms, versus service organizations, where the scope is about vertical structures and working close to the customer.

"Regardless of the balance of products and services they choose, managers of software companies must understand what their primary business is, and recognize how the two differ—for selling products requires very different organizational capabilities than selling services." (Cusumano, 2003b, p. 16)

related press release, "strengthen the focus on convergence". Whereas the previous organizational foundation was based on functional divisions, the new structure is based on a market-based approach. The corporation is divided according to its range of end-user solution groups. In addition to that, the firm is run through two complementary horizontal organizations, one of which technology-focused, and the other market- and customer-focused, which together coordinate the deliberate spill-over effects within the firm. In other words, the management of the company recognized that the maturing stage of the convergence process will not only require a growth path in the form of a technological platform, but also an *organizational platform*, which provides a capability for dynamically and rapidly launching new vertical initiatives upon.[20]

> "You now see more collaboration between business groups and business units than ever before. In order to address customer needs in this convergent environment, we need to leverage networks, key elements of the enterprise solutions, technology platforms, etc., [...] the silos are breaking down, there is more cross-collaboration and interaction between the groups. That as itself is difficult, it is a mind-shift from where [our company] has been in the past. We have been able to be successful more in our silos in the past, now it is critical as the market is evolving, to start leveraging eachother, to ensure that we're not making duplicative efforts, which is counterproductive to having an efficient, productive organization. And from a cost perspective, [...] it drives costs up, unless you really are leveraging technology throughout the company." (informant 4, case δ_2)

Commenting on the advantages of the organizational platform with respect to external collaboration, the same informant further explained:

> "What they are trying to do from a standpoint of a relationship side with our larger partners [...], is to ensure that they understand what parts of those organizations of all [our company] are interacting with these key large partners to ensure that we are getting the best umbrella agreement for a firm δ_2-at-large [...] At the same time then, each of our business groups who interact with those partners, now start understanding what eachother are doing, and we can say, 'hey,

[20] Previously, the firm had addressed convergence challenges to a smaller extent, by forming a specific convergence-group within the company, which launched and governed specific interdisciplinary projects. This first-order adaptation, however, had to leave room for the second-order adaptation, and the company abandoned that initiative.

we are looking at that as well, and what are you doing, and who are you talking with inside [partner companies], and let us work together and we will help eachother solve this problem, because we have expertise in this area, you in that area, and the 3rd party has expertise in the other area.' [...] We create this ecosystem to work together. But it is a hard paradigm shift." (informant 4, case δ_2)

In parallel to this introduction of the organizational platform, the firm δ_2, similarly to the firm δ_9, recently introduced a further second-order adaptation. In response to the platform paradigm, the company made a clear-cut step in its disentanglement of horizonal and vertical organization structures, by merging its platform-oriented network business with a major competitor, hence resulting in a separate company (where the firm δ_2 remains one of the owners).[21]

In such situations, established firms obviously need to find a match between rigid, institutionalized structures for exploiting the commercial potential within the industry, but on the other hand, being able to still change as new facets of emerging vertical mechanisms, and eventual new discontinuities, occur. In the case of firm δ_2, the decision to form an organizational platform may be regarded as the first step towards radically new organizational forms in response to the complexity of the ecosystem. Whereas technological platforms serve as a basis for rapidly developing new technological applications on top, one might ask whether firms may construct organizational platforms, which provide capabilities for dynamically producing new organizations on top of that.[22]

---

[21] Further considerations will reflect upon whether this separation along with the industry structure can be regarded as a step towards ambidexterity, splitting the company into one less innovative, first-oder adapting and rather incrementally developing, as well as one more innovative, second-order adapting and radically developing part of the company (section 4.6.3).

[22] Tushman and Rosenkopf (1992) note that in earlier stages of a technology's evolutionary cycle, organizational and interorganizational processes emerge with the goal of closing industry standards. In more mature stages of an evolutionary cycle, technological uncertainty will decrease, leaving less room for organizational dynamics. On the other hand, they also observed that the more complex the technology, the greater the amount of subsystems and corresponding interfaces, the greater the technical and contextual uncertainty, allowing room for sociopolitical and organizational dynamics to influence the technological evolution (Tushman and Rosenkopf, 1992) . Based on the observations made in this study, the phenomenon of convergence with its entropic deconstruction of vertical structures, the modularization based on horizontal platforms, and the succeeding new form of vertical reorientation, may represent a case where such organizational dynamics still matter.

In this context, the notion of *enterprise architecture* may be seen as a related recent trend for developing organizational capability in such a way, that agility and change become intrinsic, allowing a rather constructional than functional perspective to a firm (cf. Hoogervorst, 2004; Jonkers, Lankhorst, ter Doest, Arbab, Bosma, and Wieringa, 2006). Under this conceptual premise, the firm and its business model do not represent anything static, but should rather be viewed as a product of the architecture and the resources, which can be changed dynamically (Rihinen, 2006). Hence, whereas a traditional organizational view could be regarded as a frame for organizational action, an organizational architecture may represent the frame for organizational development (cf. Nadler, Gerstein, and Shaw, 1992). Whereas the firm δ_2 was observed to deliberately introduce such concepts into its organizational development practices, theory suggests that architecure-based thinking cannot be harmful even in other coevolutionary classes of maturity and evolutionary stage. After all, an organizational architecture can be regarded as a vehicle for dynamic adaptation towards contingencies that confront the firm, but has a tendency to limit the frame of action to first-order adaptation only (Ethiraj and Levinthal, 2004). In order to institutionalize the dynamics of second-order adaptation into the organizational frame, even broader architectural concepts may be required.

4.5 Reinventing the firm for convergence

Perspectives of organizational change suggest that whereas some forms of change can be predominantly influenced by structure, discontinuous shifts in organizational orientation are however generally driven by deliberate actions of executive leaders (Tushman and Romanelli, 1985). In many situations, none of the previously discussed capabilities (P_9–P_{12}) will be sufficient for managing the convergence process. Extending the argument of second-order change from the previous section, the change might even have to take place beyond the restructuring of the organizational frame. In other words, in order to fully be able to follow the cycle of convergence, firm leaders need to be clear about the intrinsic need of eventually having to reinvent the entire company. Although this need might not always become relevant in practice, a generic understanding of conceptual development paths, directions and trajectories may serve as an underlying guideline even for gradual and continuous changes through current and future cycles of convergence.

As this intrinsic need can be related to the entire process of convergence, the observations for such renewal on the level of the entire firm are distinguished between entrant and established firms. Based on the observations as elaborated in the following sections (4.5.1 and 4.5.2), a set of capability elements is identified in contingency to the coevolutionary classes. These elements are summarized in table 4.5, and yield the following proposition:

Proposition 13. (P_{13}) *Contingent on its maturity, a firm needs to develop a specific dynamic capability for a path-dependent reinvention of the value proposition and related business models along the process of convergence.*

Table 4.5: Capability elements for reinventing the firm

Entrants
out-innovate, find something new, or develop exit strategy make path dependency decide about growth path: if path fails, reinvent path, do not solely adapt path
Established
allow new metrics of the economy to reshape the metrics of the company turn path dependency into growth path: reinvent business model based on resources, not on solely adapting business model

4.5.1 Entrant firms: harness attacker's advantage

In the cases observed among the set of entrant firms, the major challenge associated to the convergent industry evolution consisted of identifying and articulating the own niche. As entrant firms usually do not have to deal with any specific inertial forces within the organizational structure or strategy vector, the freedom of choice was in many cases the basis for firms committing onto erroneous pathways from the beginning. Firms such as β_4, β_5, β_6 or α_5 started their operations based on specific business models each, which however all were more or less completely changed on the way to the present situation. It shall not be questioned that entrant firms' competitive, agile and innovative power to a large extent may be based on the nonexistent inherited roots, allowing the firms to explore their specific value creation role within an emerging ecosystem. Nevertheless, the explorative force of such a company also implies its ability to detect and accept wrong decisions, leaving room to quickly move on. For instance, the firms β_4 and β_6 both started their business with focusing on specific customer segments. After a certain time of exploration, both firms respectively realized that they were talking to the wrong customers, completely revoked previous efforts, and started from scratch, until the own niche was found.

As described above (section 4.3), the concept of mediating technologies and corresponding mediating business models can represent a highly viable strategy for leveraging complementary assets in the quickly emerging environment of spill-over-driven gaps. However, although such a mediating role may be highly profitable in the short term, firms who base their existence on such a concept have to maintain a clear awareness of future scenarios, and develop capabilities for an alternative plan. As in the case of firm β_2, such an alternative, post-convergence plan may be described by the following three generic strategic alternatives:

> "Run like hell, out in the bay, or get bought." (informant 1, case β_2)

In other words, once within a commercial, mature phase of a mediating role, one approach may be to attempt to out-innovate all other competitors—including incumbent firms who may infringe on the mediating territory, e.g., through inducing distinctive resource dependency. The second option may consist of quickly going out into the field, and searching for completely new—possibly again mediating—roles within the convergent environment, which provide the basis for reinventing the business model. Finally, the entire established business of a mediating company, including its current channels and customers, may represent a highly attractive object towards established firms. Hence, as established firms with their buying power want to enter the gap in between, exit strategies might become relevant as well, aiming at getting acquired while the mediating business still offers access to a large customer base. Admitting to be somewhat cynical, another informant concludes:

> "You need to move very quickly to become that one to get acquired." (informant 15, case β_1)

4.5.2 Established firms: rediscover assets

The convergence process can in many ways be seen as driven by a broad variety of innovative entrants, evoking continuous changes in the way technologies are applied, and in the rules of who provides what within an emerging ecosystem. Based on such disruptions, established firms might end up in situations, where the span of strategic options becomes increasingly limited. The existing strategy vector of the firm binds managerial decision-making to rather reacting to the evolutionary process, instead of actively coshaping the convergent ecosystem. Also, as initially illustrated in table 3.2, the observed strategic inflection points seem to have a stronger competence-destroying impact on established firms. This might become the case when an established firm umcompromisingly sticks to an established, historically proven business model, failing to accept the increasing amount of substituting ways of achieving same kinds of services based on previously unknown metrics. In such situations, established firms might have to rethink their origins of doing business at the very core of the company, aiming at finding new opportunities of creating competitiveness based on the combination of existing assets, as well as the new metrics of the economy.

Particularly, several telecommunication carriers may eventually need to face such fundamental reinventions. Being rigidly anchored in the middle of a well-established value chain, these firms were during the epoch of stable industry boundaries able to sustain a role of a mediator at large-scale, at both intersections of customer-to-customer (horizontal) as well service provider-to-customer (vertical). This unique and broadly tied role within the value chain has brought carriers into the privileged, and historically monopolized, position of being the sole option for the majority of customers, and thereby allowing high margins on the transactions *per se*. As the deconstruction of these vertical structures is becoming gradually but significantly altered, and the resulting competitive attacks stem from multiple directions, a strategy vector with its core consisting of the original business model may in many cases represent a dead-end street. Whereas the model at present still may be able to compete with alternative, convergence-driven business models, it does however remain 'stuck in the middle'—a middle which is not becoming anything but smaller (cf. Nyström and Hacklin, 2005). After all, as an informant puts it:

> "As an incumbent, due to deregulation, it is per definition given that you loose market share. The question rather is, how much of the 100% are you about to loose, and how many of new services models will an incumbent operator be able to pioneer or persist." (informant 29, case δ_1)

Finding themselves more or less in such scenarios, the firms δ_9 and δ_1 are in the process of recognizing the urgency of this fundamental change, implying a need for a "transformation from one company to another" (informant 5). The firm δ_9 in particular, is running an internal project aiming at reducing inertial structures and attitudes that inhibit the adoption of new economic rules into internal processes, a "change of mindset" (informant 5), which represents a necessary condition for being able to fundamentally reinvent the firm. Instead of meeting the new rules of the economy with the existing metrics of the firm, δ_9 will have to change its internal metrics based on the environment.

In doing so, the generic approach can be regarded in the reuse of existing assets instead of the existing business model. As the existing resource base of such a firm, e.g., infrastructure, install base or customer base, is based on decades of investments, it often cannot—and therefore should not—be changed. Instead, this base should be taken for granted, and should furthermore be understood as a set of rather unique critical resources, which entrant competitors hardly will be able to imitate. Hence, incumbent firms may need to find new usage scenarios for the existing resource base, allowing them to leverage a completely new kind of value proposition, which still leverages the existing base of assets—either based on these assets uniquely, or in collaboration with an external partner, turning them into complementary assets (table 4.6). An example for such an approach can be seen in the variety of carriers currently evaluating opportunities for providing payment services based on their existing infrastructures (Hacklin *et al.*, 2005b), among which also the studied firm δ_9 is.

Another interesting example for reinventing—and reincarnating—the firm can be observed in case γ_1. Throughout its history, the firm has already twice found itself in a situation, where the new rules of the economy implied traditional metrics not to work anymore, forcing the firm to reinvent itself. Originally beginning in "version 1, the DRAM company" (informant 25) developing itself into a leader of memory chip technology, the firm later fundamentally changed its business into "version 2, the microprocessor company" (informant 25). Based on its heavy investments in infrastructure and specific knowledge in the design, production and marketing of central processing units and related technologies, the firm recently reinvented itself into its third incarnation:

> "Now here we are, [company] version 3: the platform company." (informant 25, case γ_1)

Table 4.6: Resource-based reinvention of strategic approaches

Strategic approach	*Critical resource*	*Required partnerships*
Payment transaction services	Existing customer base; billing relationships; trusted, reliability brand	Financial industry; retail business
Offering high-quality data transport with end-to-end services	Network quality brand; trusted, reliability brand	Application providers
Multimedia on demand services	Large storage spaces with few "hops" from customer terminal	Content providers
Home directory services	Large storage spaces with few "hops" from customer terminal	–

Source: Hacklin *et al.* (2005b)

In other words, in the emerging spill-over between knowledge bases that firm γ_1 had stake in, as well as new areas such as communication technologies, the convergence phenomenon lead the firm's top management not only to committing to the convergent horizontal broadness through diversification, but actually to building the whole company around this idea. Hence, the firm was able to reuse its existing infrastructure for developing new technologies in synergy with new, either acquired, in-house built or even horizontally complementing partner firms, altogether providing a new, horizontal business model for complementing third party vertical applications.

> "Now convergence from [our company's] perspective has really been about embracing this opportunity as telecommunication services, data services, communications and IT converge—we see a tremendous business opportunity." (informant 25, case γ_1)

Rethinking the cores of the value proposition and corporate strategy of a firm at the very roots of an established company, might result in ways to build fundamentally new business models, through using an existing set of resources, and thereby building new, either intrafirm or interfirm, complementarities based on these. In other words, complementary assets must not only be an entrant firm's approach for quickly finding its way into niche markets. After all, even for established firms in traditional industries, complementary assets can provide a basis for reinventing the firm (cf. Hughes and Morton, 2006). The recipe may be based on a novel interpretation on existing assets, allowing the inertia of path dependency to be suddenly turned into a window of path opportunity.

4.6 Managerial commonalities

The evolutionary cycle, as determined by the stepwise augmenting impact of the convergence phenomenon, results in different environmental situations. Additionally depending on their own age and history, firms end up in different constellations with different needs for managerial capability development. However, despite the specificity of a variety of needed capabilities in accordance with the respective coevolutionary constellation, even some generic considerations can be made, independently of convergence stage or firm maturity. Whereas sections 4.1 to 4.5 elaborated on specific capabilities that are contingent to their environment, this section will discuss a few managerial issues that were observed as common among the coevolutionary classes within the study.

4.6.1 Unraveling the big picture of customer experience

When value chains deconstruct and previously linear industry structures become nonlinear, a variety of new mutual supplier, customer, competitor and hybrid roles emerge. Additionally, the succeeding new mechanisms of vertical reorientation result in complex ecosystem structures, which do not necessarily make the link between the producers and their end customers any easier. Hence, in managing the path from spill-over of knowledge bases and technologies, through applicational convergence, until the coming-together of entire industries, firms need to develop a clear view on what the end customer's role and perspective within the big picture looks like. In the end, the user does not care about the convergence itself too much, it is rather the outcome that counts.

> "Users never asked for convergence. They might ask for a single phone number, or a single voicemail box." (informant 20, case γ_4)

In particular, the customer obviously does not ask for different products or solutions with similar feature provided by different companies, he wants to experience the full convergent solution.[23] In this

[23] Similar observations were made by Edelmann *et al.* (2006).

context, the notion of customer or "user experience", reoccurred frequently during the conducted interviews. In converging environments, the customer experience is now determined by the added value of the entire ecosystem, where a single firm's solution might represent one component of such an experience only. Even though such a component might have high quality, the success of the product and thereby the providing company is measured by the performance of the synergetic solution and the growth of the entire ecosystem, not by the components *per se*. In many cases, the convergence process might not advance at all without the participation of the end-users, who are needed for allowing an emerging network effect. Hence, with an enhanced experience of the end customer in mind, firms need to move away from sticking to firm-internal incentives, towards working for the full vertical convergence needs, even though other players in the ecosystem might benefit from these initiatives as well. The existence of vertical needs are in most cases nothing new, it is the identification of mutual similarities and complementarities among them that allows a potential for bridging them together.

> "There are people that want to compute, and there are people that want to communicate. I don't think the customer cares whether he is using a computing or communication system. It's about delivering a system to the customers that meets user needs, that brings the experience they want to enjoy." (informant 25, case γ_1)

> "Originally, it was, 'let's just tell the γ_4 story', but we quickly realized that we've got to tell the γ_4 story and translate it into their vertical, subvertical, or even unique customer situation." (informant 19, case γ_4)

For creating such unique customer situations, the technological platform *per se* is not enough, a clear value proposition towards specialized vertical segments is needed. In other words, specific end-user needs cannot simply be created, or induced through a convergent application, they have to be clearly identified and understood in advance. This is what for instance the company β_6 had to learn the hard way. In the beginning, the company was all about selling a technologically superior platform.

> "[...] and our proposition was: yes, we are more expensive, and you'll probably loose your job, but hey, we're bullet proof. So we didn't meet with a very good response. [...] It wasn't until we hit upon a sort of a seminal strategy, which I call 'the primacy of caller experience'. And that is: its one thing to sell the platform. But if we do not have

commanding killer apps on the platform, nobody's going to call it." (informant 10, case β_6)

As in any business, it is the user experience that counts, but in the various complexity dimensions of the convergence phenomenon, e.g., technology, business model, partnering or competition, managerial approaches often fall short in leveraging the market pull of convergence, and rather develop a technology push. When pioneering on the synergetic effect between previously distinct knowledge bases and technologies, there are applications that might provide an enhanced experience to the end user, but there are obviously also those that just do not make any sense.

> "I guess I'm always suspicious of that notion that everything is going to collapse into one thing, because it often is a signal that engineers are beginning to drive, and that's usually a bad thing for the business. [...] I always tease my engineers with the question about why not having a converged washing machine—for both clothes and dish. [...] You can build it, yes, but chances are, that you don't want it, and might end up with something, that is less good as either, and people will reject it in the marketplace." (informant 20, case γ_4)

Induced by the observations, the following proposition is made:

Proposition 14. (P_{14}) *Along the process of convergence, a firm needs to develop a specific dynamic capability for identifying and bridging the combination of potential horizontal inter-industry spill-over with associated mutual similarities and complementarities between specific vertical customer needs.*

4.6.2 Learning the language of both worlds

As discussed in section 3.4.3, when knowledge bases and technologies converge, a major challenge in exploiting this potential can be seen in the fact that different people, with different disciplines, skill sets, languages and cultures have to suddenly work together for a common good. Before rushing into implementation, any potential for collision has to be sorted out, as collaborative innovation approaches otherwise may yield unwanted results:

> "What I've seen happen again and again is, you get these people who think they are talking about the same thing, but they're really not. You have telco people, who think in terms of switches, voice plans, minutes, reliability, redundancy, and all these kinds of things. Then you have these Internet people, who think in terms of content, IP, URLs, HTMLs, and these kinds of things. [...] People think about

> problems in very different ways. If both sides are equal, you can go around in circles for a long time. If one side is more dominant, the process may be more quickly, but you might miss important considerations. Outcomes can be things such as WAP [Wireless Application Protocol]. There would have been an Internet protocol stack, but the phone world wanted a new protocol, and the whole thing was a big failure." (informant 15, case β_1)

For avoiding such failures, the cross-disciplinary collaboration, taking place either within or beyond corporate boundaries, will have to be initially treated as a remarkable learning exercise for both sides.

> "The first step of getting them to really talk is actually to bring a common lexicon, a common language, which is behind all these things." (informant 15, case β_1)

> "We're trying to manage innovation, in terms of learning to speak the same language." (informant 21, case β_4)

Hence, when initiating development projects with the aim of either exploring or exploiting the emerging spill-over, it may be crucial to allocate a significant amount of resources and time into creating an interdisciplinary team that works. Experts need to exchange experiences, allowing them to mutually learn about history, backgrounds, and why things work they way they do. Several managers among studied firms agreed that this learning exercise has to be treated as a distinct process, that needs to be managed carefully, and by the right people. In practice, this process was often managed by engineers, which did not necessarily erode the conflicts. A rather neutral, independent project manager function might be needed, serving as an integrator between both worlds. As soon as a climate and a culture for interdisciplinary work is established, collaborative creativity within the intersection of both worlds may begin. This creativity process, however, may in some cases need to be structured as well. The project manager could lead the specification phase, aiming at agreeing on very concrete terms, i.e., what are the common requirements, what are the environmental factors between both worlds, and what are the differences.

> "[If not done that way] the problem is, you get a not-invented-here syndrome, and religious arguments of 'this is the way to do and this is not'." (informant 15, case β_1)

Hence, the argument of learning the language of both worlds can be regarded as an initial precondition for launching any convergence-related R&D project. If disciplinary and cultural differences are being taken care of initially, these investments can pave the way towards executing innovation processes with highly disruptive character.

> "Once an agreement [on the specification] is found, then it becomes just another technological development cycle. Then, either take something from the original parts and apply that, or create something entirely new. I think, too often, something entirely new is created instead of making the effort of combining existing things." (informant 15, case β_1)

Hence, the following proposition is made:

Proposition 15. (P_{15}) *Along the process of convergence, a firm needs to develop a specific dynamic capability for understanding, learning and managing the differences of involved disciplines, skill sets and cultures, as well as for integrating these into a common language.*

4.6.3 Strategic duality as an intrinsic necessity

Generally, the appropriation of an external convergence mechanism into internal value within the firm can be regarded as a major balance act. Like in any punctuated and discontinued innovation cycle, after an initial phase of enormous growth, experimentation and radical changes, firms have to find a favorable place between emerging stability and capability for next punctuations. More recent theories suggest that firms should avoid doing "too much of a good thing", arguing that a commitment to solely and simultaneously launching a broad variety of innovative products may eventually lead to failure (Barnett and Freeman, 2001). Firms should rather strive towards an equilibrium between incremental and radical innovation, since the right balance between old and new can serve as a differentiator (Katila and Ahuja, 2002). In the convergence context, there is no universal strategy for managing through this multi-level change process—an insight, which is intrinsically already contained in the contingency perspective of the analysis. But even within specific contingent situations, such as the coevolutionary classes, a general need for balancing between old and new was observed. In the shifting competitive landscape of industries coming together, firms have to find the balance between specializing in new emerging verticals, avoiding being locked into coevolutionary obsolescence, and on the other hand making use of existing capabilities—creating something else out of them. Especially during later periods of an evolutionary cycle, firms need to develop the ability to produce incremental innovation even as they develop competencies to develop subsequent technological breakthroughs (Anderson and Tushman, 1990). This requires firms to allow a certain type of strategic stretch within their capability development, an approach that puts a premium on the ability to develop multiple, often inconsistent competencies simultaneously (Burgelman, 1983). In particular, firms should not only distinguish between exploration and exploitation in their organizational learning activities, but also find and maintain a balance between them (Benner and Tushman, 2003; Brown and Eisenhardt, 1997; Levinthal and March, 1993; March, 1991, 1996). After all, an exclusive managerial focus on the development of new capabilities may as an outcome induce dysfunctional forces and rigidities that inhibit innovation (Leonard-Barton, 1992).

The managerial challenge of this strategic duality, consisting of a mental balancing act between constantly looking forward and backward, implies a need for firms to develop capabilities for ambidexterity (O'Reilly and Tushman, 2004; Tushman and O'Reilly, 1996). Ambidextrous organizations have the ability to create value based on dichotomous characteristics, such as

- pursuing both revolutionary and evolutionary change (Tushman and O'Reilly, 1996),
- simultaneously exploring and exploiting innovations (Benner and Tushman, 2003; March, 1991),
- creating and sustaining advantages (Grant, 1996),
- implementing responsiveness and efficiency (Hanssen-Bauer and Snow, 1996), as well as,
- undertaking change and preservation (Volberda, 1996).

Further reflecting the previously elaborated observations regarding industry change and managerial capabilities to theory, one can assume that regardless of firm size, the approach of bipartite organizational design represents a way of achieving ambidextrous organizations, allowing horizontal and vertical initiatives to be separated within the company (section 4.4). On the one hand, the platform concept represents a feasible approach for managing through convergence processes, as a platform has the ability to capture horizontal spill-over potential across established industry boundaries (section 4.1). On the other hand, however, platforms do not survive without their complementors (Gawer, 2000; Gawer and Cusumano, 2002), implying a need for creating an environment, where third party providers are incentivized to complement. Hence, the organization must be able to master horizontal and vertical development of innovative activities at the same time, which in some observations lead to the impetus of—at least partly—splitting firms into according bipartite organizations.[24]

For instance, both firms within the class of reincarnating giants, the firm δ_3 pursues an innovation approach that strives towards multiple efforts for developing similar things in parallel, allowing a later comparison of which way actually works best. Then again, in the firm δ_2 the current strategic effort is dealing with reducing redundancy between the innovation processes in business groups, as knowledge bases have become together. Apparently, whereas the firm δ_3 in this sense currently might prioritize towards rather explorative efforts, and firm δ_2 towards exploiting ones, both sides may need to be maintained in parallel. Similarly, speaking for the entrant firm α_5, the informant commented:

> "One of the things that we have to be very careful about is to not eat our own dogfood. On the one hand, we have to laser focus on execution and market opportunities. But we also have to recognize that we may not be right and have to be nimble enough to change things." (informant 17, case α_5)

Exogenous activities of a firm are equally challenged by the causal dichotomy, of on the one hand adopting partnering activities in response to the environment, on the other hand thereby imposing structural consequences to the environment. In particular, collaborating horizontally may on the one hand create new value to the end-user based on convergent applications, but is the other hand associated with risks to the firm. The coevolution of firms can on the one hand be characterized through horizontal collaboration, aiming at diversifying towards previously unrelated industries and achieving platform leadership. By doing so, on the other hand, the horizontal collaboration across industry boundaries may namely induce the horizontal spill-overs, thereby accelerating the convergence process. As a result, mutual horizontal and friendly collaboration can suddenly turn into coopetition.

The innovation strategy of the established firm δ_6 reflects this duality around the issue of collaboration:[25]

---

24 In this context, the work by Jansen (2005) presents empirical evidence for firm ambidexterity being more positively related to firm performance when exploratory and exploitative innovations are separated in different organizational units, than when these are combined in organizational units.

25 In fact, Cohen and Levinthal (1990) indicate that investment into in-house R&D activities increases the absorptive capacity of a firm, i.e., the ability to gain innovation from outside the company.

> "[...] the way [our company] is doing it: 'Ok, I'm going to, in some sense, parallel process, I will develop it in-house, but I'm going to be very willing to look outside and acquire and partner with other companies'." (informant 16, case δ_6)

Hence, the following proposition is formulated:

Proposition 16. (P_{16}) *Along the process of convergence, a firm needs to develop a specific dynamic capability for managerial ambidexterity, continuously balancing between exploring and exploiting current and emerging opportunities of inter-industry spill-over.*

Capabilities of strategic duality and ambidexterity were observed to be crucial in dealing with the idiosyncratic dynamics of convergence. However, the goal shall not be to create an environment to allow dissonance within strategic approaches (Burgelman and Grove, 1996), but rather the ability to understand the clear distinction between short-term and long-term strategies, or pre-convergence and post-convergence policies.

5

Managing through cycles of convergence

Anderson and Tushman (1991) note that top management has a tendency to pay attention to industry recessions, along with a willingness to make painful cost-cutting moves when demand drops. However, it is not that form of competition which threatens the survival of firms and its rivals. It is argued that above all technological change, not downturns in demand, are associated with shake-outs. This requires a need for maintaining the organization's ability to navigate through cycles of technological change, characterized by technological discontinuities and implied creative destruction. After all, managerial efforts for directing the firm's marketing and financial operations provide a solely incremental instrument for improving profitability. On the contrary, the capability to ride waves of product and process innovation affect not only the profitability, but in a more long-term perspective, the viability of the entire firm (Anderson and Tushman, 1991).

5.1 Deriving managerial guidelines

Whereas the previous chapter emphasized on identifying and articulating a set of inherent capabilities that provide a basis for accurate managerial action, this section will focus on the interpretation of these insights. Building on the approach used by Wasson (1974) and Hofer (1975) in formulating appropriate strategies over the product life cycle, it is in this section attempted to formulate a contingent management model based on the evolutionary perspective. A summary of propositions is given in table 5.1. As the set of propositions consists of both contingency-related capabilities, as well as managerial commonalities, it is further necessary to distinguish between both perspectives in deriving and formulating firm-level managerial guidelines.

Table 5.1: Summary of propositions

Identifier	*Proposition*	*Page*
Evolutionary perspective (sections 3.2–3.5)		
P_1	The phenomenon of convergence can be comprehended as an evolutionary process of technological change, which is initiated through coevolutionary spill-over between knowledge bases of distinct industries, and can successively expand into more applied levels of convergence, which eventually can lead to the merging of entire industries.	65
P_2	Along the process of convergence, a cautious transition from exploration to exploitation of knowledge is needed. Whereas exploration activities might lead to novel competencies on the one hand, these may at the same time induce inertial forces for exploitation.	76
P_3	Along the process of convergence, an opening-up of proprietary innovation processes is needed.	81
P_4	As knowledge bases and higher level applications converge, involved firms' disciplinary horizons need to diverge.	83
P_5	Along the process of convergence, vertically integrated structures of value creation disaggregate into fragmented horizontal value creation systems.	90
P_6	Along the process of convergence, collaborative dynamics of specialization and verticalization emerge, in response to fragmented horizontal value creation structures as suggested in P_5.	94
P_7	Along the process of convergence, the value creation opportunities for intersectional applications develop from mass to niche markets.	97
P_8	The process of convergence undergoes a cycle of successive increasing and decreasing competition, which is successively punctuated by the emergence of local and eventually global dominant designs.	100
Contingent capabilities (sections 4.1–4.5)		
P_9	Contingent on its coevolutionary class, a firm needs to develop a specific dynamic capability for understanding, developing and maintaining a platform advantage for capturing the horizontal spill-over between knowledge bases and its higher-level applications.	113
P_{10}	Contingent on its coevolutionary class, a firm needs to develop a specific dynamic capability for understanding, anticipating and orchestrating an ecosystem of coopetition and reciprocal incentives.	124
P_{11}	Contingent on its coevolutionary class, a firm needs to develop a specific dynamic capability for exploring and exploiting complementary assets that leverage the inter-industry spill-over, as well as for anticipating and reacting on the limited lifetime of such assets.	138
P_{12}	Contingent on its stage within the convergence process, an established firm needs to develop a specific dynamic capability for internalizing structural industry changes into its organization through first-order or second-order adaptation.	144

Table 5.1: (continued)

Identifier	*Proposition*	*Page*
P_{13}	Contingent on its maturity, a firm needs to develop a specific dynamic capability for a path-dependent reinvention of the value proposition and related business models along the process of convergence.	151
Non-contingent capabilities (sections 4.6.1–4.6.3)		
P_{14}	Along the process of convergence, a firm needs to develop a specific dynamic capability for identifying and bridging the combination of potential horizontal inter-industry spill-over with associated mutual similarities and complementarities between specific vertical customer needs.	158
P_{15}	Along the process of convergence, a firm needs to develop a specific dynamic capability for understanding, learning and managing the differences of involved disciplines, skill sets and cultures, as well as for integrating these into a common language.	160
P_{16}	Along the process of convergence, a firm needs to develop a specific dynamic capability for managerial ambidexterity, continuously balancing between exploring and exploiting current and emerging opportunities of inter-industry spill-over.	163

As the proposed dynamic capabilities in propositions P_9 to P_{13} are in contingent relationship to a firm's respective coevolutionary class, these cannot be simply compiled into a general view. In order to achieve a holistic view of the induced theoretical framework, a further clustering is needed. The projection of each capability onto a respective coevolutionary class can yield a basis for gaining a consolidated view related to each class. This consolidated view aims at suggesting guidelines for managerial capability development throughout the evolutionary cycle of convergence.

The associated elements of these capabilities (as articulated in tables 4.1 to 4.5) have been categorized based on the type of dynamic capability, i.e., distinctive organizational processes, a firm's specific assets positions, and the coevolutionary path it has adopted or inherited.[1] Based on this categorization, the elements of identified dy-

[1] In particular Teece *et al.* (1997) suggest that the essence of dynamic capabilities and competitive advantage of a firm rests on the three dimensions of processes, positions and paths. The *process* dimension of a dynamic capability is based on distinctive organizational and managerial processes, i.e., ways of getting things done through coordinating, integrating, combining or learning. In particular, the dimension describes properties on a firm's ability to reconfigure and transform. The dimension of *position* refers to a firms specific assets, that represent a basis for competitive advantage. These can consist of technological, knowledge-based,

namic capabilities (i.e., P_9–P_{13}) have been rearranged into a coevolutionary class-oriented perspective. Table 5.2 contains the summarized juxtaposition of contingent elementary capabilities, distinguishing between process, position and path-oriented type. Based on this conceptual management model, the following firm-level recommendations for scenarios corresponding to the coevolutionary classes can be made. Further translating and applying the elementary capabilities into a conceptual summary, table 5.3 attempts to distill managerial paradigms on the level of processes, positions, and paths, respectively.

complementary, financial, reputational, structural, institutional as well as market assets. Finally, the *path* dimension captures the idea of a firm's position and paths ahead being a function of previous paths. Hence, the match between endogenous path dependency and exogenous technological opportunities determines a firm's ability to develop itself into new directions (Teece *et al.*, 1997).

Table 5.2: Contingent elementary capabilities for managing through convergence

Category	**Early stage***		**Late stage****	
Entrant firms	*Pioneering disruptors*	*Prop.*	*Vertical attackers*	*Prop.*
Processes	publish and explore open platform, allow others to build on top of that	P_9	identify existing technological and business model conflicts and construct platform based on that	P_9
	stick to open standards	P_9	from supplying established firms to competing with them	P_{10}
	anticipate emerging ecosystem	P_{10}	out-innovate, find something new, or develop exit strategy	P_{13}
	grow into new rules of competition	P_{10}		
	establish heterogeneous set of external relationships	P_{11}		
	out-innovate, find something new, or develop exit strategy	P_{13}		
Positions	leverage neutrality advantage	P_{10}	disrupt ecosystem, break rules	P_{10}
	find complementarity niche	P_{11}	induce established firms' dependency on own assets	P_{10}
	focus on inimitability of assets	P_{11}	induce dependency and lock-in of established firms towards mediating model	P_{11}
	mediate between established business models	P_{11}		
Paths	create clear differentiator for platform	P_9	create growth path based on convergent platform	P_9
	explore new technologies for potential platform construction	P_9	exploit new technologies; leverage spill-overs between downstream layers	P_9
	anticipate emerging conflicts between established business models	P_{11}	post-convergence path; further distinguish complementarity or exit	P_{11}
	make path dependency decide about growth path: if path fails, reinvent path, do not solely adapt path	P_{13}	protect niche advantage; avoid becoming obsolete	P_{11}
			make path dependency decide about growth path: if path fails, reinvent path, do not solely adapt path	P_{13}

Table 5.2: (continued)

Established firms	*Platform consolidators*	*Prop.*	*Reincarnating giants*	*Prop.*
Processes	acquire external competencies for creating new complementarities	P_9	acquire external competencies for upgrading and substituting internal competencies	P_9
	open-up innovation process to smaller players	P_9	'chinese wall', sharp distinction between own and shared intellectual property	P_{10}
	aligning innovative activities into emergence of ecosystem	P_{10}	harness ecosystem for launching new innovation cycles	P_{10}
	co-shaping new rules of coopetition	P_{10}	build margins on (new) combination of assets not on assets per se	P_{11}
	leave vertical integration behind	P_{10}	exploit structures of organizational convergence; rebuild organizational entities based on industry evolution	P_{12}
	value network assimilation	P_{11}	market-driven adaptation of organizational processes	P_{12}
	explore ways of creating organizational convergence; try new organizational combinations	P_{12}	second-order adaptation	P_{12}
	first-order and second-order adaptation	P_{12}	allow new metrics of the economy to reshape the metrics of the company	P_{13}
	technology-driven adaptation of organizational processes	P_{12}		
	allow new metrics of the economy to reshape the metrics of the company	P_{13}		
Positions	align existing technological assets into platform, spanning converging space	P_9	identify out-of-the-box platform opportunities among existing resources	P_9
	capability platform thinking (technology, knowledge, partnerships, organization)	P_9	position oneself as experienced, reliable partner towards entrant firms in ecosystem	P_{10}
	develop total platform advantage	P_9	value network orchestration; induce incentive and lock-in of entrant firms to participate in own innovation	P_{11}
	develop scale advantage; make too costly to imitate	P_{11}		
	find complementarity dominance	P_{11}		

Table 5.2: (continued)

Paths	explore complementary assets	P_9	exploit complementary assets	P_9
	identification of internal R&D synergy potential	P_{12}	protect dominance advantage; avoid becoming commoditized	P_{11}
	turn path dependency into growth path: reinvent business model based on resources, not on solely adapting business model	P_{13}	identification of internal and external commercial synergy potential	P_{12}
			turn path dependency into growth path: reinvent business model based on resources, not on solely adapting business model	P_{13}

* Knowledge and technological convergence
** Applicational and industrial convergence

Table 5.3: Contingent managerial paradigms

Category	**Early stage***		**Late stage****	
Entrant firms	*Pioneering disruptors*	*Prop.*	*Vertical attackers*	*Prop.*
Processes	shape and syndicate mechanisms of collective action	P_9–P_{11},P_{13}	infringe on upstream and downstream layers of value generation	P_9,P_{10},P_{13}
Positions	build inimitability through neutral mediating with complementary assets	P_{10}, P_{11}	induce and abuse established firms' resource dependency	P_{10}, P_{11}
Paths	scale horizontally between and beyond industry boundaries	P_9,P_{11},P_{13}	achieve dominance, specialize upstream, or exit	P_9,P_{11},P_{13}
Established firms	*Platform consolidators*		*Reincarnating giants*	
Processes	search for external complementarities and invest in mechanisms of collective action	P_9–P_{13}	search for internal complementarities through permutation of contradictory assets	P_9–P_{13}
Positions	build platform leadership and induce complementary business models	P_9, P_{11}	leverage reputation and dominance for ecosystem orchestration	P_9–P_{11}
Paths	grow horizontally beyond industry boundaries, refine and replace business models	P_9,P_{12},P_{13}	protect dominance, grow upstream, or reinvent core value proposition	P_9,P_{11}–P_{13}

* Knowledge and technological convergence
** Applicational and industrial convergence

5.1.1 Pioneering disruptors: shape, build, scale

Being an entrant firm at the beginning of an emerging convergence cycle, the very cores of the value proposition and associated business models need to build on the capability to possibly well understand and anticipate emerging trajectories, and thereby emerging potentials for disrupting previously established industry environments.

As the technological uncertainty at this stage of the evolutionary cycle is high, it is crucial for such a firm to find a way of riding the technological trajectory, as it usually cannot afford multiple directions of resource commitments. One way to do so is based on working with open standards, aiming at representing a complementary, but distinctive asset within an emerging platform-based ecosystem. As neither the technological, nor the market uncertainty allows for too strong resource commitments onto specific vertical customer segments, the development of platform capabilities can allow the company to grow along the broadness of convergence, leveraging the horizontal spill-over effects between industries, and thereby sustaining a growth path for emerging verticals. Hence, the organization of internal processes should center around leveraging emerging exogenous effects, particularly involving inter-firm syndication of the new mechanisms for collective action.

Also, in order to be well positioned in early phases of value chain deconstruction due to convergent spill-overs, the entrant firm may construct internal development, sales and partnering processes to follow the emerging rules of the new ecosystem, allowing it to collaborate with a heterogeneous set of firms (including competitors), through leveraging a neutrality advantage. This should represent the basis for building the firm's inimitability, as the neutrality advantage secures the firm's position in mediating and thereby providing complementary assets.

Being an entrant firm, the impact of path dependency can generally be regarded as low, at least compared to the competing incumbent firms. Therefore, strategic opportunities should be aligned along with the process of convergence, rather than as a reaction to coevolutionary mechanisms. Being in the position of pioneering a possibly disruptive technological cycle, the path should be aligned along scaling according to whatever new spill-over effects during the convergence evolution in an agile and dynamic way, allowing the firm to hold a path of pioneering and disrupting.

5.1.2 Vertical attackers: infringe, induce, achieve

Being an entrant firm at mature stages of the convergence process, capability development needs to capitalize on the head-start it has compared to its established firm rivals, allowing the firm to ride the trajectory opportunities, rather than to react to them (chapter 4, figure 4.2 and definitions 14–15). In particular, a firm of such coevolutionary constellation may on the one hand search horizontally, i.e., through identifying existing technological and business model-oriented conflicts allowing them to capture horizontal spill-over effects through platform-based approaches. On the other hand, the probably even more significant advantage of such a positioning is based on the combination of low path dependency and tangible changes in the industry structure. As the competitive constellations are becoming completely rebuilt, complementary assets can turn into substitutes, and the mediating role can quickly become an outperforming one. Instead of horizontally mediating between established business models' conflicts, the entrant firm may develop various windows of opportunity for infringing these vertically.

Hence, the firm may gradually evolve from supplying established firms to competing with them. In other words, the firm may develop a position of not only inducing, but also abusing their resource dependency. Through on the one hand inducing a lock-in of established firms towards the mediating model, the act of coevolutionary rule breaking may not only disrupt the ecosystem, but can on the longer term even yield the lock-out of established players.

Instead of solely acting in a mediating role, the firm has to base its growth path on the convergent platform, thereby creating capabilities of exploiting the horizontal advantage under the new mechanisms of verticalization. However, if chances for further specializing or exploiting the complementarities seem low, the mediating business model might have reached its limits and shall be subject to redefinition.

5.1.3 Platform consolidators: search, build, grow

As an established firm at the early stages of the convergence process, capability development can be characterized by the reactive position towards the evolutionary trajectory (chapter 4, figure 4.2 and definitions 14–15). In terms of evolutionary timing, this type of firm would per definition be at least in sync with convergence cycle. Hence, capabilities for transforming the dominant position into actively co-shaping the evolutionary mechanism are needed. In particular, the proprietary innovation system needs to be opened-up towards integration of entrant players' activities, allowing the established firm to align its existing activities within emerging decentralized innovation structures, allowing the firm to learn and participate within the new mechanisms of collective action. By doing so, the firm will be well positioned to capture the combined advantage of being both an experienced, resourceful company, as well as finding itself in the early stages of the convergence process. As knowledge bases and technologies start to generate horizontal spill-over effects, investment into collective innovation models will allow the firm to match external complementarities with internal asset portfolios.

In particular, this setting proposes the aim of achieving platform leadership as a beneficial position. By opening-up closed innovation models well in time and establishing horizontal linkages based on both proprietary and external assets, the building-up of horizontal dominance seems by far more feasible than attempting to retain vertical integration. The platform advantage may be distinctive through the firms internal complementary assets, allowing the development of scale benefits, which become too costly to imitate from an entrant perspective. Instead, this position can induce incentives towards entrants for rather developing complementary business models instead of imitating, which further strengthens the platform advantage. In other words, an established firm's ability to achieve platform leadership may induce coevolutionary lock-in from the perspective of complementary entrant firms.

A longer-term perspective in such a setting would suggest the firm to continue its expansion and growth on the horizontal dimension, i.e., prioritizing rather diversification than specialization of its knowledge base. Based on the horizontal leadership positioning, the path should consist of identifying and capturing further horizontal spill-overs between industries, as the firm might be best in place to capture the full convergence broadness advantage compared to its coevolutionary competitors. Hence, for as long as possible, the firm should attempt to gradually refine, before entirely replacing its business models.

5.1.4 Reincarnating giants: search, leverage, protect

At mature stages of the convergence process, established firms might need to initiate deeply rooted processes of change, if not already done so during earlier stages. As indicated by the nomenclature of this coevolutionary class, the firms might need to create a reincarnation of the entire company. Since the convergence process is in its commercial and thereby highly exploitative phases, there is not much room for novelty left in such a setting, leaving the firms' established business models endangered towards attacks from entrant firms, who in turn may outinnovate and outperform based on the new metrics of the ecosystem. Also, at this stage, a late-follower strategy of entirely opening-up innovation processes may yield fatal results, as this may even strengthen the risk of business models being attacked. Hence, processes of reinventing the business model should prioritize on rediscovering the resources of the company, before fatefully searching for external complementary assets. The search should mostly disregard existing business models, but rather create capabilities around new resource combinations that may gain completely novel ones.

In particular, the inherited industrial dominance, although based on outdated metrics, may represent a position for exploiting reputation and scale advantages in the scope-oriented stage of the convergence process. By creating new and powerful resource complementarities, the firm can create capabilities for orchestrating the mature ecosystem, inducing incentives towards particularly entrant firms. These entrant firms would on the one hand be attracted by the existing customer base of the firm and by the associated opportunities for scaling, which may even allow the established firm to generate lock-in effects. On the other hand, as the established firm will have to focus on vertical niches, an orchestrated participation of a broad set of entrant firms will allow the model to both scale and scope, i.e., through serving a broad portfolio of vertical market needs.

In other words, the path orientation of the firm should create capabilities for protecting and sustaining the dominance, through adaptation towards the new metrics of the ecosystem. This will enable the firm to grow upstream, delivering solutions to the emerging vertical reorientation mechanisms. In many cases, however, this transition may not happen over night. For reincarnating a 'giant' firm, it may be necessary to reinvent at the resource-based cores of the value proposition.

5.2 Cyclical determinants of convergence

Given the fact that firms are likely to go through more than single stages of the convergence process, it is implied that not only situational and contingent managerial actions for capability development have to be taken. But even in the longer term, the anticipation of scenarios for more mature phases of the convergence process shall represent part of managerial capability development. Hence, the capability development process can be depicted as a transition between levels of dynamic capability in the short term, as well as new aspects of contingent requirements in the long term. Figure 5.1 captures both dimensions of the capability development process for both entrant and established firms.[2]

Hence, this sequential perspective of capability development based on situational contingencies suggests guidelines for managing a firm through the cycle of convergence. Being based on a dynamic resource-based view of the firm (cf. Helfat and Peteraf, 2003), these guidelines articulate elements for understanding and anticipating the contingent requirements on dynamic capability over time. Another managerial issue, however, is based on the question what happens thereafter. Based on the observed cases, there are indications that firms do not necessarily only experience one single convergence cycle during their lifetime. For instance, this can be seen as the case in several firms whose business models are either closely or distantly associated to cellphones. Whereas the first convergence cycle happened around the spill-over between communication platforms, i.e., Internet technologies and established circuit-switched telecommunication lines, the next cycle can already be anticipated. Through the pervasive diffusion of digital media solutions, new spill-overs around the digital camera, TV, music and other forms of entertainment emerge. Hence, by having the digital still camera integrated into cellphones, industry boundaries are increasingly becoming blurred, and the digital camera industry might sooner or later be forced to react. Drawing on this insight, one may infer that the currently observed convergence phenomenon between previously distinct areas of telecommunication, information technology, media and entertainment, can be regarded as a series of convergence cycles. In other words, instead of viewing the evolutionary development as driven through an initial serendipitous spill-over between several knowledge bases at once, the development can be understood as a series of multiple local convergence cycles that build on each-other (figure 5.2).

5.2.1 Structural reiteration

Theories of industrial change suggest the reocurrence of regular patterns, which firms shall anticipate, align their development life cy-

[2] Obviously, this distinction assumes managerial opportunism instead of reactionism (chapter 4, figure 4.2, and definitions 14–15), aiming at rapidly developing along the convergence process. On the other hand, in the case of trajectory reactionism, where firms mature faster than the convergence process evolves, the long-term capability development direction in figure 5.1 would be represented by the migration from entrant firm to established firm paradigm in each recommended action item (i.e., the transition from white to black, rather than from left to right).

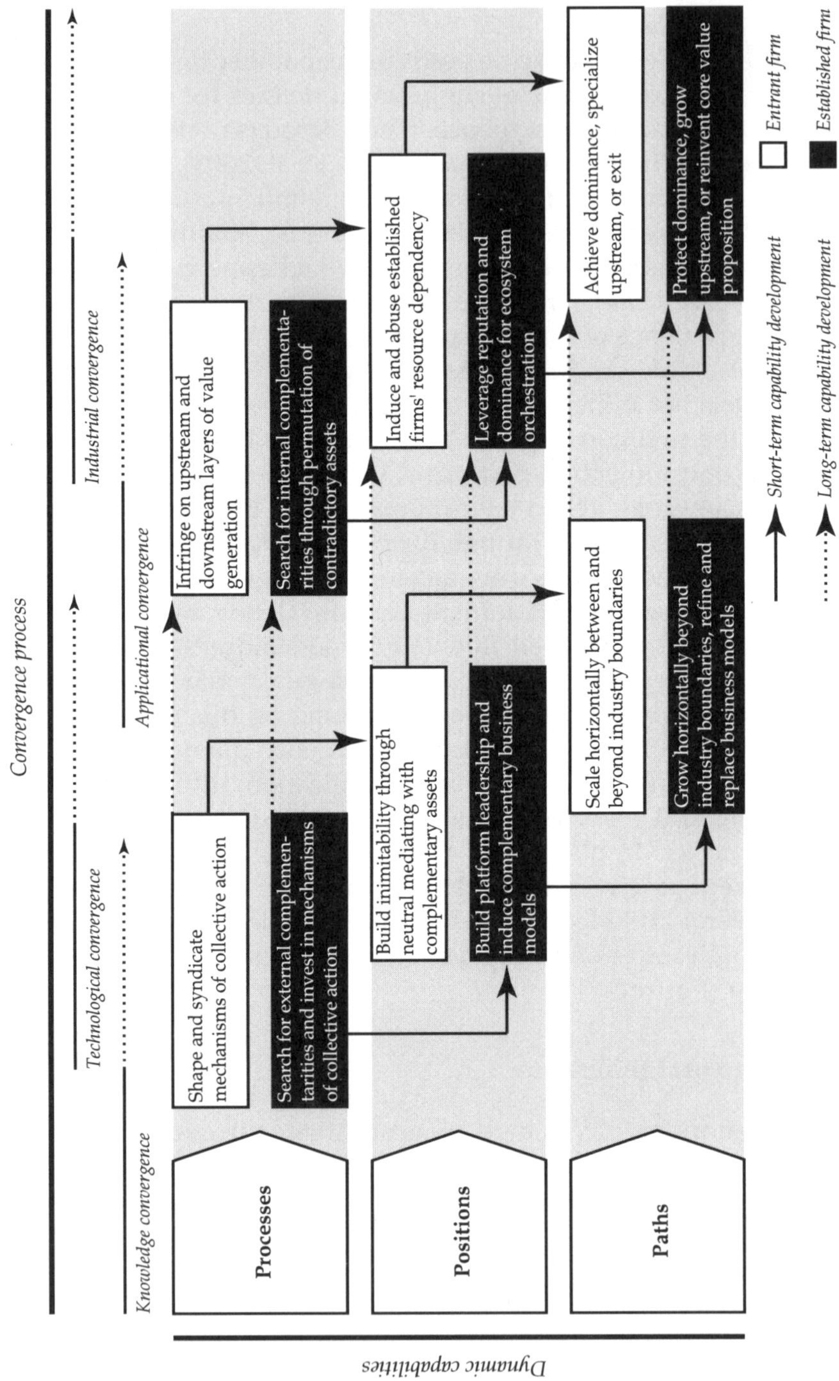

Fig. 5.1: Guidelines for the capability development process

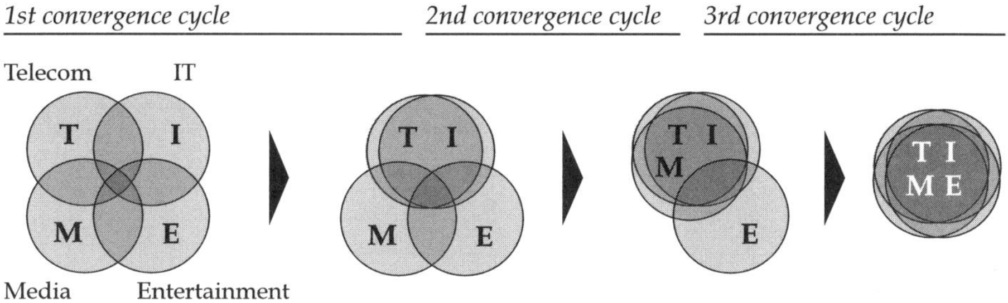

Fig. 5.2: Cyclical perspective on observed convergence

cle with, and translate into their organizational evolution (cf. Anderson and Tushman, 1991; Fine, 1998; Klepper, 1997; Tushman and Romanelli, 1985).

In particular, Tushman and Rosenkopf (1992) suggest the cyclical reiteration of phases of variation, fermentation, selection and retention (section 2.3.1, figure 2.2). In the context of convergence, an increased deconstruction and entropy of technology-driven systems, followed by the formation of new, higher-level vertical mechanisms, was observed along these phases. However, along the repetition of this cycle, Tushman and Rosenkopf (1992) furthermore infer an occurrence of multiple levels of openness of the technological system. As the number of dimensions of merit tends to increase with every evolutionary cycle, technological systems can be regarded as communities, that transform from nonassembled products, to open systems, along several punctuations and emerging dominant designs. Hence, with every newly initiated evolutionary cycle, the level of complexity further increases.

Applying this insight to a convergence perspective, one can assume that in certain situations, the end of one convergence process may represent the beginning of another one. Elements of the technological system, that previously originated from different industries and came together based on mutual spill-over, may in the next convergence cycle represent one single entity, that can be embedded into further mechanisms of recombination (figure 5.2). Making reference to the formalization as introduced in section 3.2.5, figure 5.3 illus-

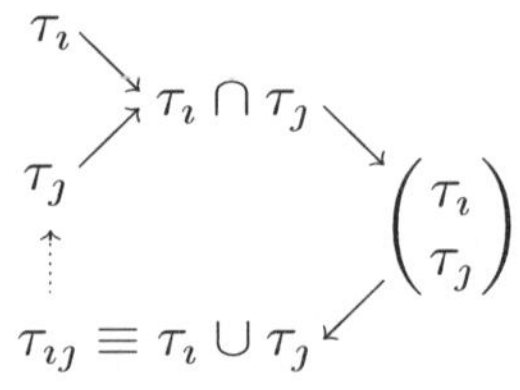

Fig. 5.3: Elementary perspective on evolutionary cycles of convergence

trates the elementary mechanism of embedded technological complexity from a cyclical convergence perspective.[3]

Hence, assuming that the need of continuous organizational adaptation to the reoccurrence of structural change cycles also holds for the convergence phenomenon, the previously induced management model may need to be considered as a recursive process as well.[4] Translating this to the coevolutionary perspective, the capability lifecycle shall be based on multiple cycles of coevolution. As the end of one convergence process can be the beginning of another one, from an entrant firm's perspective, the capability development process has to be based on the assumption, that along with multiple evolutionary cycles, not only does the contingent environment coevolve, but so does the company itself in terms of age and maturity. Again assuming opportunistic managerial behavior, entrant firms in both early or late stages of the convergence process (i.e., pioneering disruptors and vertical attackers, respectively) may mature and experience the next cycle from a platform consolidator perspective. For established firms, in turn, the transition into new convergence cycles will imply an organizational punctuation and managerial response. Figure 5.4 depicts the cyclical implications on the contingent management model.

[3] For similar considerations on cumulative embeddedness of cyclical systems, cf. Hacklin, Lopperi, Bergman, and Marxt (2004b).

[4] In examining the exogenous impetus to the organizational perspective, Tushman and Romanelli (1985) stress the alternation between cycles of organizationally convergent periods "which elaborate structures, systems, controls, and resources toward increased coalignment", as well as mechanisms of reorientation, which consist of "periods of discontinuous change where strategies, power, structure, and systems are fundamentally transformed towards a new basis of alignment" (Tushman and Romanelli, 1985, p. 173).

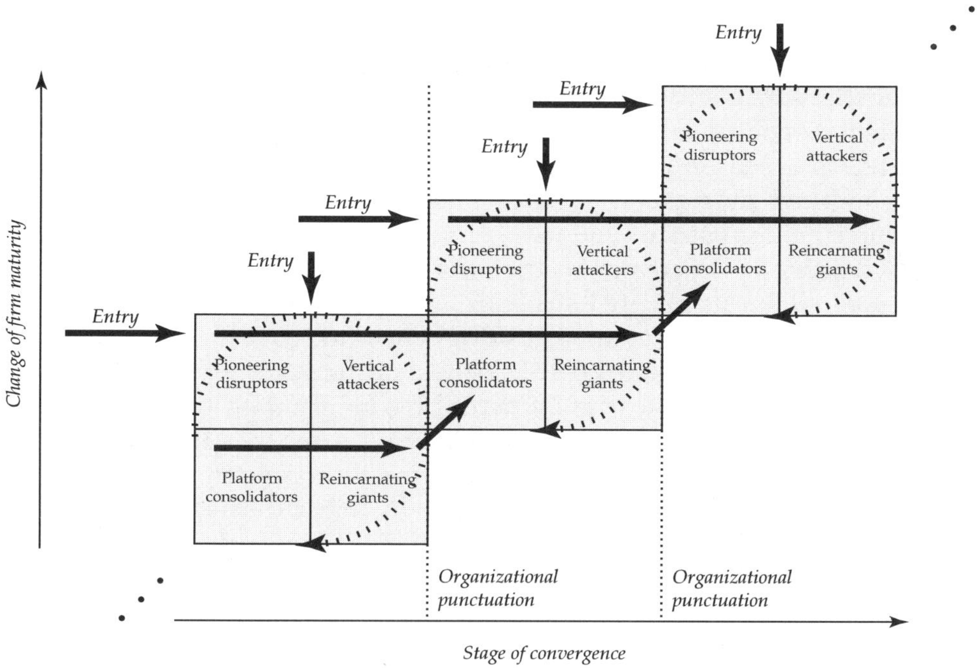

Fig. 5.4: Convergence as cycles of coevolutionary change

5.2.2 Distinctiveness of convergent cycles

Fine (1998) suggests the industry evolution to alternately cycle between vertically integrated and vertically disintegrated structures. Similar structural effects could be observed in the context of convergence. In this study, it has been observed how vertically integrated industries start to disintegrate, followed by a punctuation as caused by the sudden spill-over between knowledge bases of distinct industries. As same tasks can be performed by players outside the integrated value chain, the vertical structure is disaggregated. New players enter the value creation system, aiming at connecting the horizontal gaps. Hence, platformization and by that, horizontalization of the industry structure occurs over time. However, as the convergence process reaches maturity and thereby commercial stages, new specific end-customer needs have to be addressed, implying a need for customization and introduction of new mechanisms for vertical specialization.

Hence, the competitive edge turns vertical again, and new efforts of vertical integration emerge.

However, when reiterating this cycle, one can identify specific characteristics of the convergence process, which differ from the way usual recursive evolutionary processes might happen. In particular, the difference is based on the fact, that the structural industry change process implied by the convergence phenomenon does not take place within rigid industry boundaries. As industry boundaries change, the convergence cycle creates restrictions for to what extent the industry may regain its original structure. In other words, a discontinuity under rigid industry boundaries originates from *within* the industry, and can have e.g., technological, societal or regulatory sources. Based on this disrupted state, the industrial reconfiguration tends to reorganize into new linear structures, allowing firms to reintegrate vertically through acquiring vertical margins. In a convergent discontinuity, however, the source for disruption stems from the spill-over of knowledge bases *between* industries, making it rather impossible for the industry to reestablish its vertical structure. Hence, nonlinear industry structures occur, making vertical reintegration no longer a viable solution from a firm perspective. Table 5.4 highlights the key differences between structural change in changing and rigid industry boundaries.

Further emphasizing on the differential role of entrant and established firms in this context, figures 5.5 and 5.6 illustrate the differences between structural industry change cycles in rigid and changing industry boundaries. In particular, figure 5.5 depicts the transition from an initially vertically integrated state (1) into a disintegrated state (2) where entrant firms infringe both horizontally and vertically on incumbents' business models, creating a temporarily nonlinear state. However, as cores of value generation tend to remain alike, a vertical structure is partly recreated, where both established and entrant firms reorganize themselves vertically into the consolidating state (3). Finally, as the industry cycle matures and consolidation entails the shake-out of firms, a vertically integrated structure is retained, resulting in the stabilizing state (4). In contrast, the cycle in figure 5.6 commences from an initial state of two vertically integrated value chains (1). The occurrence of an inter-industry spill-over yields a transition into the disintegrating state (2), where not only entrant firms infringe on existing margins, but also established firms may be able to diversify horizontally or specialize vertically. The consolidating state (3) implies the new ecosystem to emerge where new rules and metrics shape non-

Table 5.4: Structural dynamics of industry boundaries

	Industry boundaries	
	Rigid	*Changing*
Initial state	Single vertically integrated industry	Multiple vertically integrated industries
Characteristics of discontinuity	Within industry Core competences become core rigidities Outperformed vertically Pressure to disintegrate	Between industries Convergence spill-over between industries Same tasks can be performed by actors outside of the industry Pressure to disintegrate
Implication on industry	Modularization Share margins with others	Value chain deconstruction Share business with other industry
Managerial action	Modules become commoditized Incentives for vertical reintegration occur Acquire margins of buyers and suppliers Recreate vertical structure	Vertical integration no longer a viable solution due to resource constraints Rather platform approach, remain within own level of integration Collaborate vertically, best to exploit platforms in other areas than acquiring buyers or suppliers Vertical co-specialization
Result	Vertically integrated structures reoccur	Emergence of vertical market segments, but no vertical integration

linear structures towards vertical specialization. Finally, as the system starts to stabilize, consolidation takes place between industry boundaries, e.g., causing established firms from previously distinct industries to merge, or even leading them to shake-out (4).[5]

As opposed to the case of rigid industry boundaries, the transition into the initial state within changing industry boundaries is hence associated with some limitations. For further convergence cycles, the initial state is most probable to be characterized by less linearity, as the ecosystem may not have clear vertical structures. In further cycles of convergence, it may hence be assumed that it is much less visible

[5] In figure 5.6, one may observe semantic linkages between the states (1) to (4), as well as the convergence stage definitions 3 to 6 (sections 3.2.1–3.2.4).

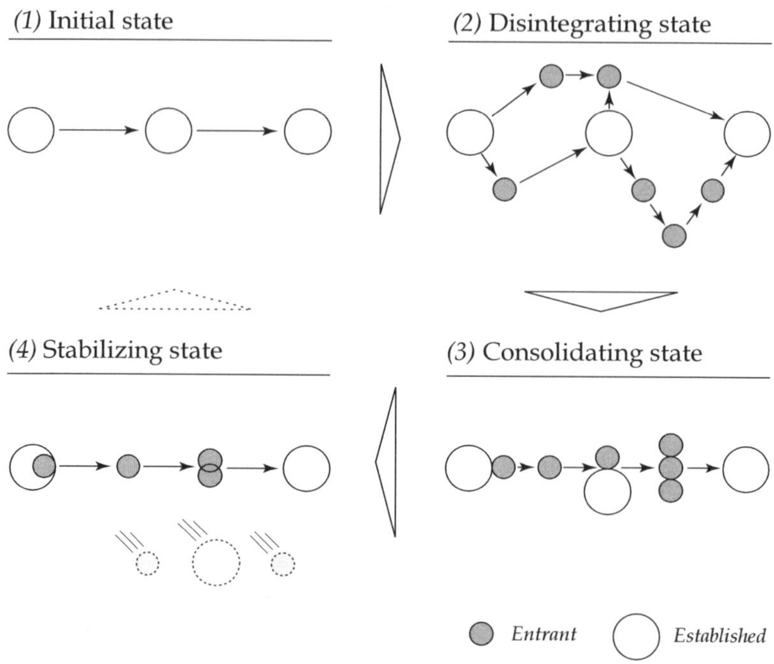

Fig. 5.5: Cyclical change under rigid industry boundaries (discontinuity)

where to directly infringe business models, as value generation mechanisms have become complex and highly embedded.

Apart from structural industry evolution and cyclical repetition of change processes, another cyclical peculiarity of the convergence process can be seen in its recursive causality, which goes back to the coevolutionary interplay of search and selection (sections 3.1.2 and 3.3.1). In particular, as firms react on the convergence phenomenon through managerial action, the collective action additionally tends to further reinforce the evolutionary process (figure 5.7). In particular, initiated by the horizontal spill-over between knowledge bases of distinct industries (0), firms need to respond by diversifying horizontally. Thereby, firms collaborate horizontally across industries in order to embrace the convergence potential (1). As firms start doing so, accompanied by the entry of new firms, the spill-over effects further increase, as firms from both distinct industries come closer together in terms of what they know and what they do (2). This further accelerates the convergence between industries, causing further vertical deconstruction

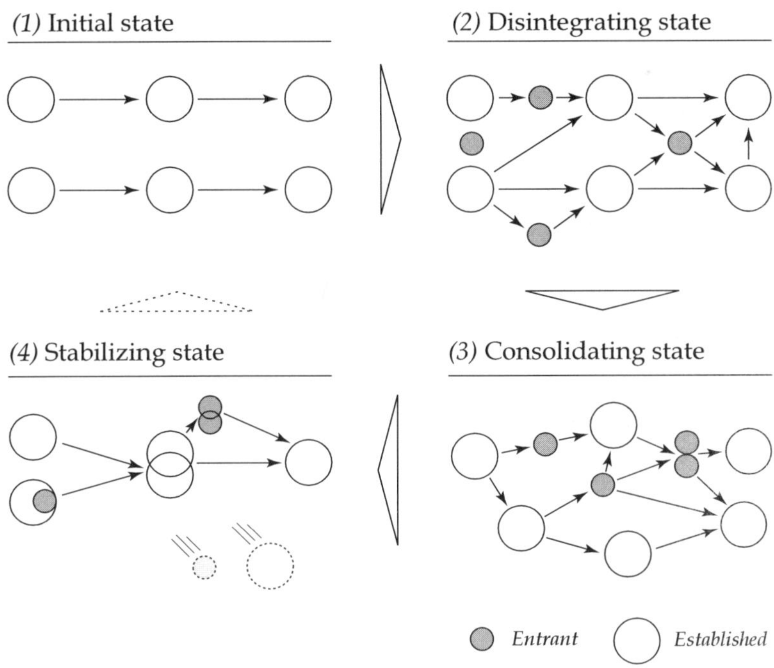

Fig. 5.6: Cyclical change under changing industry boundaries (convergence)

and horizontal competion, as firms through the coming-together not only have become more similar, but thereby also mutual competitors (3). Hence, managerial efforts for mastering 'coopetition' is needed, implying firms to both collaborate and compete, both horizontally and vertically (4). In order to exploit the convergence potential, and deliver added value to the end-user that differs from previously available products or solutions, new vertical perspectives occur, forcing firms to collaborate horizontally. By acting without reciprocal incentives, it might not be possible to get the platform layers to work, which however is needed to leverage the full user experience (5). This, in turn, leads back to the first step (1).

5.3 Transfer and application of the cyclical model

5.3.1 From retrospective to predictive

Previously described observations regarding structural industry inflections, firm-level challenges, and managerial capability develop-

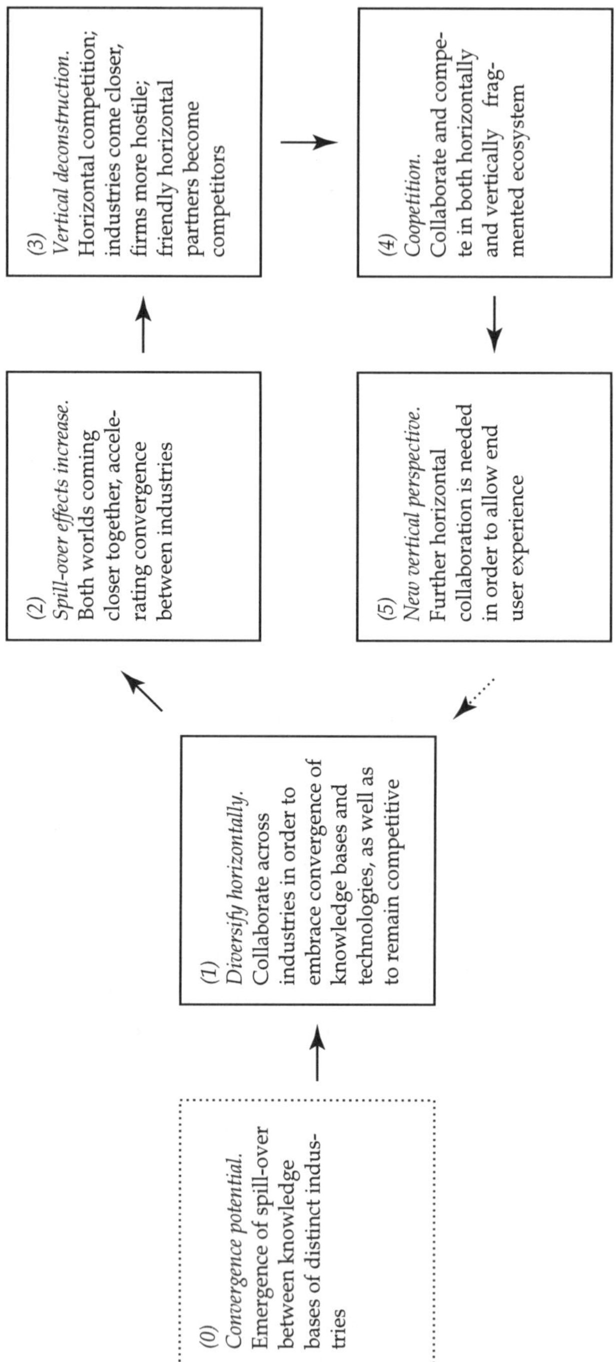

Fig. 5.7: Causality cycle of convergent industry change

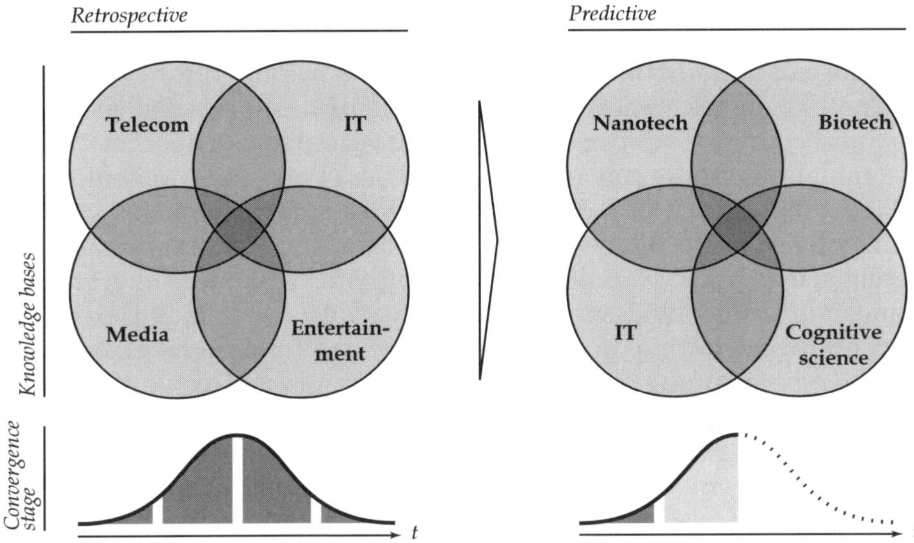

Fig. 5.8: Extrapolating logic of model transfer

ment formed the basis for inducing theory and deriving models. As these considerations are based on the sequential characteristic of the convergence process, the induced theory considers patterns of the phenomenon on both firm and industry level on a retrospective basis, i.e., *ex post*. As the underlying concept of comparative replication among the studied case objects was based on the coevolutionary development of firms and industries, the model may allow learning and recommendations for emerging convergence cycles, i.e., *ex ante*. Obviously, emerging convergence cycles can only test the model to a limited extent, as the data of more mature phases still is missing. Instead, the induced theory derived from the retrospective view may serve as basis for anticipating emerging patterns of the selection mechanism, as well as developing scenarios for predicting new forms of search. Hence, emerging cycles of convergence do not allow any replication of the analysis, but an analysis of the extrapolation of previously generated propositions into the future. Figure 5.8 depicts the logic of model transfer applied to the second set of empirical analysis within this study.

5.3.2 NBIC as a next convergence cycle

The increasing amount of spill-over effects within the intersection of the disciplines nanoscience, biotechnology, information technology and cognitive science (NBIC) can be regarded as an indicator for the initiation of new convergence processes (Roco and Montemagno, 2004). Whereas the key driver for creating spill-over effects in ICT convergence could be seen in the increasing degree of digitalization, recent scientific developments around nanotechnologies suggest that nanoscale science will act as a similar catalyst for entailing convergence between technologies, and for acting as a fundamental platform technology spanning a wide range of existing scientific and techno-scientific fields (Meyer and Davis, 2003; Roco and Bainbridge, 2002a,b, 2005; Tegart, 2002). The concept of nanotechnology can be comprehended as the emergence of "materials and systems whose structures and components exhibit novel and significantly improved physical, chemical and biological properties, phenomena and processes due to their nanoscale size" (Tegart, 2002, p. 2). The purpose of the discipline is to harness these properties by achieving control of structures and devices at atomic, molecular and supramolecular levels, which in turn provides a basis for learning how to efficiently manufacture and use applications based on these. Instead of simply introducing new applications, nanotechnology is widely regarded as a new paradigm, bringing the industry to a threshold of a revolution in the ways which materials and products are created (Tegart, 2002). In particular, the scientific field of nanotechnology results from an underlying knowledge convergence of traditional fields of chemistry, physics, mathematics, biology and engineering sciences (figure 5.9).

Hence, nanotechnology shall not be regarded as an industry, but rather as a phenomenon of existing industries, as they converge on an atomic scale (Kleinberg, 2005). An informant described it the following way:

> "Nanotechnology as a whole I don't view at all as an industry. It is a way of doing things, it is a scale in which one can work, it's an enabling technology that is obviously going to be, it is applied to a dozens of different industries, so as such it is kind of hard to characterize. It's unique and potentially a very advantageous way to do something. So if one views it that way, rather than an industry, as more an enabling technology, then the problem in characterizing it boils down in simple facts, to which industry are you talking about, because in each industry, the challenges are consequently different." (informant 40)

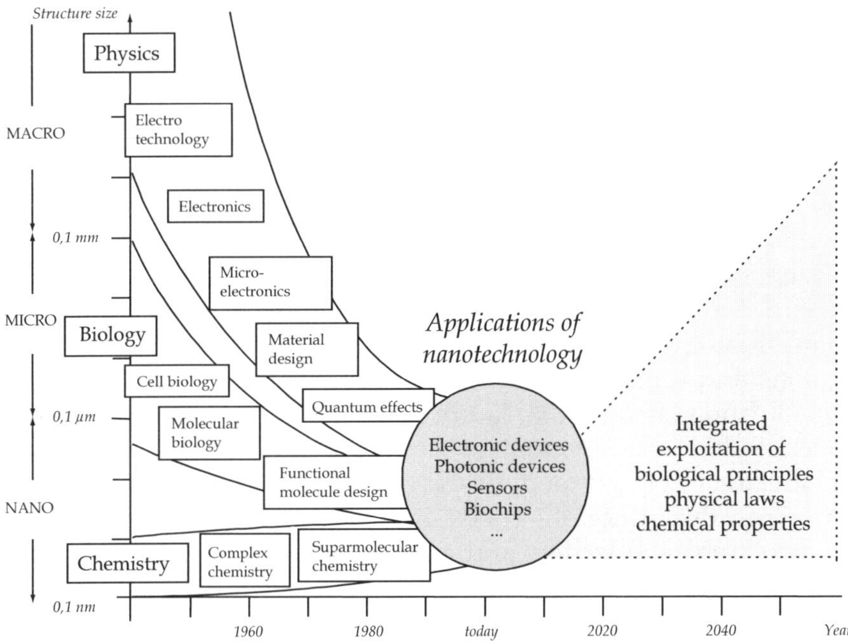

Source: Bachmann (1996)

Fig. 5.9: Trajectories of underlying scientific disciplines

However, similarly to previously observed implications, it is suggested that convergence around nanotechnologies may have the potential to disrupt existing industries, implying a need for anticipatory measures (Roco and Bainbridge, 2005). The scope of disruption may be manifold:

> "The daunting challenge of managing rapid and complex technological-driven change is increasingly a disruptive force on today's markets, business, economics, and society. Disruptions will cut more deeply as innovations fostered by convergent technologies emerge more quickly. At the same time, new opportunities will offer unprecedented market leadership for those prepared to exploit them." (Roco and Bainbridge, 2002b, p. 72)

Advances in science and engineering on a nanoscale level can launch and sustain economic progress: nanotechnology may reduce input costs in some industries, at the same time improving productivity in others. Also, it will shift demands from certain goods to others (Roco and Bainbridge, 2005).

The spill-over effect between knowledge bases of distinct industries may initiate the emergence of hybrid technologies and new applications at intersectional areas. For instance, the intersection of nanotechnology and ICT may result in nanodevices, nanosensors and other areas of nanoelectronics. Similarly, the combination of nanotechnology with biotechnological systems can yield bioelectronic applications, nanowires, nanotubes, microfluidics, or advances in drug delivery. The convergence of all three areas, i.e., ICT, biotechnology and nanotechnology provides a basis for the development of biosensor solutions and biological nanoscale chips (cf. Kleinberg, 2005; Rojas-Chapana and Giersig, 2006, figure 5.9).[6]

Hence, there is a potential for nanotechnology-related convergence mechanisms to be initiated. Based on the second part of data collection within this study, the focus of the following sections will be to find structural similarities and transfer learnings about managerial patterns to NBIC convergence.

5.3.3 Industry-level implications

Applying proposition P_1 upon this context, one may anticipate the stepwise augmentation of the nanotechnology-driven convergence evolution, initiating higher-level spill-over effects, thereby changing rules of roles of associated business models and shifting industry boundaries. Hence, as absurd as similar ideas on competitive changes might have sounded ten years ago from the perspective of IT and telecommunications industries, NBIC convergence allows speculation for e.g., pharmaceutical and electronics companies eventually becoming mutual competitors. Although to a much lesser extent, observations going into this direction were made during the interviews. For instance the case of firm λ_5 (section 5.3.4), which as a company focuses on nanoscale solutions, experiences increasing competition from biotechnology firms. To cope with this trend, along with proposition P_3 and P_4, an increasing significance of multidisciplinary intra-firm and inter-firm collaboration, thereby opening-up innovation mechanisms, was observed among experts' and entrepreneurs' perceptions.

[6] In fact, the emerging development of NBIC convergence embeds previous advances in ICT developments, i.e., the outcomes of a previous convergence cycle. This strengthens the argument on increasing system complexity along with cyclical iterations, as elaborated in section 5.2.1 and as depicted in figure 5.3.

"One will need to be flexible and fast in this environment. Big pharma is neither, so I think big pharma will seize to exist as such a dominant player as we see them today. It is not going to happen overnight, because they have so much money. I think big pharma will end up spinning off different units, to deal with more specialized areas and in a smaller entity and then compete successfully. Biotech companies, I think, are in the right general space already, they are smaller, faster and very flexible. So I think you will see big pharma come down and biotech companies rise up to a similar level. This will become of a much more knowledge driven approach, a customized approach, than the 'Ford Motor company way of doing'. Big pharma today is like the Ford Motor company, one car after another with a very small number of customizable options." (informant 31)

Being in early stages of the convergence process, the spill-over between NBIC-based knowledge bases and the induced coming-together of technologies shows structural tendencies as suggested through propositions P_5 and P_9. As horizontal spill-overs do not only lead to changed competitive constellations, but furthermore entail vertically integrated industry structures to be disaggregated, one can anticipate trends of horizontalization in previously established innovation structures, such as e.g., drug development. Hence, as biotechnology and nanotechnology-oriented companies may deliver similar or better solutions than established pharmaceutical firms, they might horizontally infringe on established business models. In order to capture the broadness of horizontal spill-overs, i.e., the multitude of disciplines associated to NBIC, horizontal collaboration is needed, and a vertically integrated drug delivery model may disaggregate into a horizontal technology-driven open ecosystem.

However, although a deconstructed and horizontalized industry structure may enhance the disciplinary broadness and amount of possible solutions involving convergent spill-over, the technology-driven industry structure will have to develop abilities for targeting specific customer needs (P_6 and P_{14}). In particular, what can be regarded as one emerging applicational issue in the realm of NBIC convergence, is represented by the concept of individualized medicine.

Advances in biomedical sciences have contributed to a wider, but at the same time more detailed understanding on what makes the body work. For this reason, current trends pave the way for a trend of specific therapy becoming tailored to an individual, meaning that customers do not necessarily need to go to the drug store and buy pills at same strengths in the future. Instead, using all of these different techniques, the physician will determine what targeting sequences are necessary to get a drug exactly where it needs to go inside the body. This sort of individualized and specialized therapy is not that sort of business which the large pharmaceutical companies can capitalize on—entrant players may be better positioned here. Hence, new industrial structures for balancing between scope and scale, exploration and exploitation, internal and external R&D will be needed.

> "[...] if you think about the personal computer and you assume that it is a drug, so you have the circuit board and you have a standard micro processor [...]. Maybe that's the nano part of this. There is going to be some new engine for this to which people will attach their favourite hard drives, their favorite graphic cards, etc. So, the nanogroups may come in in that sense, and each pharmaceutical company then will attach an antibody or therapeutic to it and then sell it to the pharmacy, to the hospital, or even directly to the patient." (informant 31)

Hence, the intersection of nanotechnology-based delivery mechanisms and pharmaceutical applications may represent a viable scenario for industrial horizontalization and vertical co-specialization. Based on a disintegrated, but still vertically oriented industry structure, a vertical modularity of products and solutions can be facilitated (figure 5.10), which in turn corresponds to propositions P_6 and P_7. Nanopharmaceutical platforms may emerge, providing the convergent basis. This could for instance allow pharmaceutical companies to deliver customized, vertical solutions, based on a horizontal system. An overview of such an industry-level transition is depicted in figure 5.11.

> "I think this is a little bit different in the sense since it is entirely multidisciplinary, the platforms will be difficult to define, because it is connected to so many different areas, and of course you will have many platforms in all sort of areas. As time goes by, the platforms will develop but it is hard to define those. I think the gene is already out of the pocket and nobody can put it back in, its just too late, so the best thing is to manage it." (informant 33)

5.3.4 Firm-level recommendations

Based on the induced propositions P_9 to P_{13}, managerial action has to be taken under the premise of the specific environment, which is contingent upon firm maturity and the stage of the convergence process. Obviously, as the NBIC convergence process can be regarded as within its early stages, where knowledge bases and technological

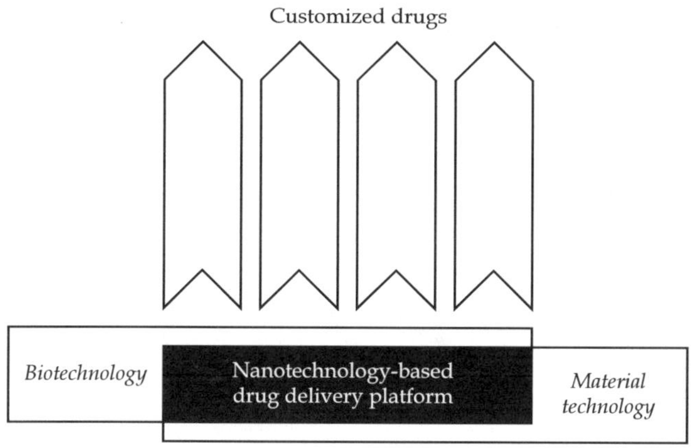

Fig. 5.10: Targeted delivery through platform-oriented verticalization

domains of previously distinct industries come together, this setting omits the availability of data for the coevolutionary classes of vertical attackers and reincarnating giants. Apart from making reference to the anticipative propositions P_1 to P_8 regarding strategic inflection points along the evolutionary mechanism, recommendations will be based on transferring contingent managerial guidelines from the coevolutionary classes of pioneering disruptors, as well as platform consolidators. As this part of the case study is focused on expert and stakeholder interviews, rather than on replicative logic, it is chosen to feature two case companies, from of which one is an entrant and one is established.

Entrant firm: case λ_5

Founded in 2000, the US-based firm λ_5 focuses on nano-electronic detection solutions, maintaining a portfolio of devices based on one common core technology. These scaleable devices use ultra-sensitive carbon nanotube detection elements combined with proprietary chemistries, and can be deployed across a broad range of industrial and medical applications where valuable attributes, such as low power consumption, small size, and high sensitivity, offer significant performance advantages, at the same time enabling unprecedented access to critical information.

In particular, the firm has product lines in three different categories: industrial detection devices, medical breath analysis devices, and bio-

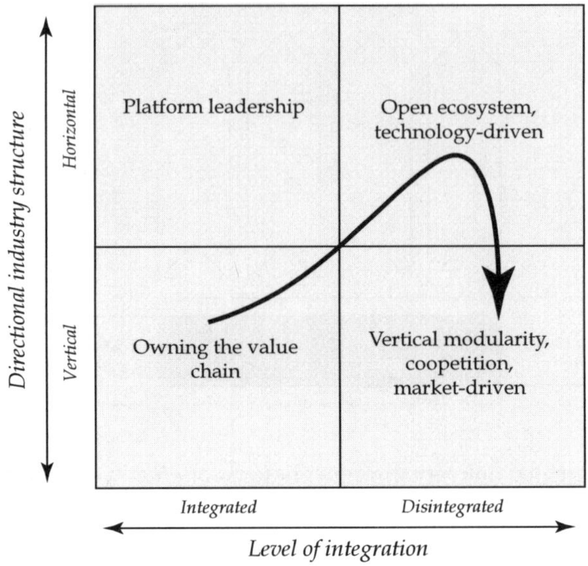

Partly adapted from Vesa (2006)

Fig. 5.11: Industry-level transition through several configurations

detection devices. These products target specific customer needs, such as immediate access to critical information as a basis for decision making (e.g., portable detection of dangerous gases, viruses in breath, or emergency respiratory measurements), a growing demand for medical point-of-care information, the emerging promise of genetics and proteomics[7] to diagnose and treat diseases, or the demand to simplify complex testing protocols making them practical for routine use (e.g., clinical diagnostics, point-of-care diagnostics, drug discovery, human identification, personalized therapy, continuous monitoring, or biomedical research).

Currently, the corporate management is in the process of partly refining the vision and the identity of the company, trying not to think of themselves as a "nanotech" company, but rather using this to really develop specific products and applications, where the nanoscale technology provides real competitive advantage. Hence, being a technol-

[7] Proteomics denotes the large-scale study of proteins, particularly with regard to their structures and functions. The nomenclature establishes an analogy with genomics.

ogy oriented organization, the firm strives towards developing itself into a more market- and application-driven firm.

> "And so the challenges that we face are—certainly there is always the technological challenge—doing something that is really application specific. So we have got a detection platform, where we put carbon nanotubes to a variety of substrates and we turn them into sensors, and we functionalize them while we put different chemistry recognition technologies onto the tubes to create these specific sensors. So there are challenges, technical challenges, in making these sensors work in very different applications, very different environments, very different performance ranges." (informant 40)

However, apart from these rather technology-specific, and life-cycle challenges, another major current problem was perceived in the implications of the platform concept. Based on the technological broadness implied by the platform approach, where the company has developed core capabilities for capturing horizontal spill-over effects and integrating these into a distinctive synergy-levaraging system, there is a discrepancy between market scope targeted by the platform and firm-internal resource constraints.

> "[It's about] separating the very promising and attractive opportunities from the distractions and the things that don't really pay. They are not the things that we should be putting our time and energy into. So deciding where to go and what to do, we put a lot of energy to that over the last nine months or so. And in fact, the company has been all along, but in terms of really granting up that effort we have really put a lot of focus and attention to being comfortable and confident in saying 'no' to a lot of things." (informant 40)

Hence, as the company on the one hand wants to leverage the full broadness of convergent added value, on the other hand however cannot, one may suggest the company to look for emerging changes not only in terms of industrial boundaries, but also in terms of changing and emerging rules of the ecosystem. Being an early mover from the NBIC convergence process perspective, the firm is well positioned to actively shape and syndicate mechanisms of collective action. Hence, the firm may need to further develop processes that allow their platform solution to be viewed and understood as a complementary asset within the industry—also from incumbents' perspectives. By doing so on a broad basis, the firm's inimitability could rest on the distinctive complementary asset, further benefiting from the firm's neutrality advantage—especially when dealing with established companies from e.g., medical industry sectors, who according to the informant tend to be difficult in collaborating with due to their dominant positions.

> "The platform is fundamental to our value. And it is critical that we have a platform, because you don't have to put your eggs in one basket. We have many different channels and many different things, that can bring revenue, value and growth for our company. And given that, you don't have to have one thing to be the absolute huge success, now the chances are like anything, if you have five interesting product concepts, one or two could be great, one or two might be average and one could be less than great. It's like any portfolio, but that is good, that is what makes us an attractive company. We can have an average product and it can add to the value. If all you have is one product and it is average, you are in trouble. If you have a platform and a portfolio, it evens out." (informant 40)

Nevertheless, in terms of the path dimension of capability development, it may be recommended to focus on extending the platform in its full broadness, creating and sustaining a growth path beyond specialization. Hence, instead of solely administrating and specializing the product portfolio based on the core platform, investments into further horizontal scaling—beyond industry boundaries—may pay off in the future. This is particularly the case, as soon as the commercializing and specializing convergence trajectory allows established players to on the one hand come closer to the cores of the business, and on the other hand easily imitate too specific service models.

Established firm: case λ_4

Founded 1987, the US-based firm λ_4 is a biopharmaceutical company that focuses on discovering, developing and commercializing therapeutics, striving to advance the care of patients by developing medicines that make a significant difference in the treatment of life-threatening infectious diseases. Although less than 20 years old, the firm can still be considered as rather mature and experienced compared to its entrant competitors. The firm's current portfolio consists of less than ten commercially available products. Furthermore, the firm focuses its research and clinical programs on anti-infectives.

Targeted customer groups consist of physicians on the one hand, as well as patients and caregivers on the other. The firm provides physicians with resources to better care for patients by rapidly developing innovative therapeutics. Furthermore, the firm collaborates closely with patient advocates, designs drugs to enhance treatment regimes, and implements patient assistance programs to ensure the firm's efforts help meet the needs of patients and their caregivers.

Among other products, the company is selling a mechanism for drug delivery using lipid spheres of 100 nm in diameter that encapsulate a specific drug. While primarily a biotechnology company, the firm represents an example of the emerging use of nanotechnology in the applied bioscience, such as the pharmaceutical sector. Hence, the firm's identity is in the process of transition.

> "[The firm] λ_4 is a drug company. [...] So we are kind of a very strong pharmaceutical company. But in someway this company is managed like a biotech company, so sometimes this company is called a biotech company." (informant 36)

Altough the company possesses a variety of established R&D mechanisms, the nanotechnology dimension could be perceived as in a rather explorative phase. Internally, the company is in the process of overcoming inertial disciplinary structures, aiming at building new internal mechanisms for multidisciplinary collaboration.

> "[...] when you are doing drug development you have a wide range of folks involved with that because you need to, it's a little different because we are a molecular bio company, we are not making an end product, we are not making an end-body, we are not making a protein." (informant 36)

But apart from searching for complementarities inside the organization, clear tendencies for addressing the horizontal spill-over potential could be observed in the way of dealing with the outside world. Hence, where firm-internal competencies are not enough in order to address new trends of nanotechnology, the firm is in the process of learning to orchestrate and eventually integrate the surrounding ecosystem of smaller players.

> "But I am sure a lot of companies have launched programs or collaborations [...] they want to know what companies like [us] are interested in. What essence would be useful to λ_4 because then λ_4 will buy them. And buy their system." (informant 36)

Hence, the tendencies of investing into mechanisms of collective action could be observed. However, apart from solely screening and eventually acquiring for externally available competencies, a beneficial strategy for the firm would be to further invest on the platform approach, than screening for external knowledge based on specific vertical demands only. In particular, the build-up and acquisition of a broad horizontal knowledge base, spanning current and emerging horizons of convergence should preferably be envisioned. By aiming at a position of platform leadership, and thereby inducing complementary business models, the firm would not need to endlessly acquire firms and their competencies based on emerging verticals, which would not be a real scalable and sustainable approach after all. Apart from cost-related constraints regarding scaling, the clarity of the firm's strategy vector may suffer, and the company may become difficult to maneuver.[8]

[8] In terms of future paths of the company, there was not sufficiently data available to compare with previously generated constructs and propositions. Hence, apart from the generic paradigm of growing horizontally beyond industry boundaries, aiming at consolidating internal complementary assets into a platform, and if necessary partly refining core business models based on a resource assessment, no further specific recommendations can be formulated.

6

Conclusions

> "And he that will not apply new remedies must expect new evils; for time is the greatest innovator. [...] It were good therefore that men in their innovations would follow the example of time itself; which indeed innovateth greatly, but quietly, by degrees scarce to be perceived. For otherwise, whatsoever is new is unlooked for; and ever it mends some, and pairs other; and he that is holpen takes it for a fortune, and thanks the time; and he that is hurt, for a wrong, and imputeth it to the author."
>
> *Francis Bacon (1561–1626)*[1]

As observed throughout this study, the phenomenon of convergence can occur in different facets with different implications on firms and industries. The variety of these facets is on the one hand determined by the evolutionary stages, as the further the process of change advances, the broader are scope and impact of the phenomenon. The phenomenon may occur in rather conceptual stages, where knowledge bases become blurred, but industry boundaries still remain unchanged. As the convergence process advances, the impact moves upstream along the value chain, and may eventually yield a changing nature of previously established industries. On the other hand, the specific impact on individual firms is determined by the respective coevolutionary setting. Not only does stage and maturity of the industry-wide change phenomenon *per se* matter, but the particular size, age, and inertial heritage of an organization at specific stages of the process represent the basis for contingent capability development and strategy formulation. An entrant firm may find itself with very different strengths, weaknesses, opportunities and risks with regard to a specific convergence stage, than an established firm. Hence, the

[1] From essay XXIV "Of Innovations", reprinted in Bacon (1909)

Table 6.1: Research questions and propositions

Research questions	*Propositions**
Q_1 Are there key capabilities which enhance the management through processes of technological convergence?	P_9-P_{16}
Q_2 How can the dynamics of convergence be comprehended from an evolutionary perspective?	P_1
Q_3 How does technological convergence influence innovation management practice?	P_2-P_8

* For an overview of propositions, see table 5.1 (p. 166)

dimension of time seems to occur as a crucial parameter of the convergence phenomenon in two ways. On the one hand, it determines the stage of the convergence process, and on the other, the maturity of firms with respect to that process. In either way, the fundamental evolutionary mechanism is affected, as conditions for search and selection vary depending on respective mutual constellation of industry-wide process and firm-specific configuration.

6.1 Discussion of results

6.1.1 Key findings

Evolutionary perspective. Aiming at answering subquestion Q_2, evolutionary perspectives on the convergence phenomenon were investigated (chapter 3). Based on comparing the observation of firms with different specific challenges regarding the management of convergence effects, four different stages of convergence were identified, described and formalized. It is suggested that the phenomenon of convergence proceeds along an evolutionary trajectory, representing a foundation for determining the dynamics of innovation (proposition P_1, cf. table 6.1).

Coevolutionary dependency. Aiming at answering subquestion Q_3, implications of the convergence phenomenon on innovation management practices within firms were analyzed (chapter 3). Building on the evolutionary perspective, the strategic positioning of firms in converging environments was identified as dependent upon the respective coevolutionary class. Hence, the observation of how single firms coevolve with the convergence process as well as with other firms, represented the basis for understanding and comparing changes and challenges for managerial practice (propositions P_2-P_8, cf. table 6.1).

Contingency of key capabilities. By answering both subquestions Q_2 and Q_3, the elaboration on managerial capabilities aimed at answering research question Q_1 (chapter 4). Further building on the observed coevolutionary dependency, key capabilities for managing through the convergence process were identified and described in contingency to respective coevolutionary classes. Hence, depending on size, age and inertial heritage of a firm, as well as depending on respective phase along the convergence process, the required set of capabilities for managing through convergence differs (propositions P_9-P_{16}, cf. table 6.1).

Managerial guidelines for capability development. Further aiming at answering research question Q_1, the set of identified capabilities for contingent coevolutionary classes was translated into a procedural view on capability development (chapter 5). Based on such a conception, managerial guidelines for managing through convergence processes were identified, articulated and summarized for each coevolutionary class. Recognizing a relationship between the convergence phenomenon and cyclical behavior of technological change cycles, the elaborated guidelines were applied and exemplified for an emerging case of convergent change.

6.1.2 Contribution to theory

In previous research, the phenomenon of convergence was rather analyzed in single situational observations, with a lack of an integrated or longitudinal understanding. Based on reviewing the literature, a generic understanding on the state of research on the convergence phenomenon was generated, and previous conceptions by various scholars were disentangled into antecedents and implications. In particular, the review yielded contradictions among existing perceptions. By introducing the evolutionary perspective on convergence, the phenomenon does on the one hand present new empirical evidence suggesting convergence as a special form of technological change. On the other hand, the elaborated process from knowledge convergence to industrial convergence (definitions 3-6) integrates previously distinct fragments from theory into a holistic evolutionary understanding.

Drawing on underlying theories from evolutionary economics,[2] the notion of convergence as a multi-level phenomenon is extended. In particular, the initial conception by Lei (2000) is extended, in considering the phenomenon of convergence on both firm and industry-level, with the respective association to mechanisms of search and selection. It is furthermore inferred that convergence can be regarded as a coevolutionary phenomenon, as single firms do not only coevolve with the surrounding population of firms, but also along with the convergence process. Building on such a classification, it is further argued that firm-level analysis of the convergence phenomenon needs to take place under the premise of contingency theory. As the coevolutionary setting of each single firm is dependent on its contingent environment, this context intrinsically impedes the formulation of any universal guideline.

Hence, through laying the theoretical foundation of the analysis on an evolutionary economics perspective, the succeeding iteration between empirical observation and generation of new theory was consequently built on the notions of search and selection. By pursuing such a multi-level approach, the inductive observation of reoccurring aspects, as well as the resulting formulation of propositions, established contextual relationships between the convergence phenomenon and a variety of literature streams. In particular, models generated within this study address various areas of management theory, i.e., technological change and trajectories, organizational evolution, industry evolution, population ecology of firms, contingency theory, as well as, management of strategic change.

6.1.3 Implications for practice

From a practitioner's perspective, this work puts emphasis on the notion of convergence as a process, rather than a momentary external trend. In particular, along with the evolution of the convergence process, the impact tends to gradually increase in scope and disruptiveness. Hence, understanding the phenomenon and dealing with its antecedents and implications requires a great deal of foresight in innovation and strategic management practices. After all, the role of top management consists of rather recognizing transitions than initiating them (Burgelman, 1994). For doing so, the evolutionary perspective and the resulting model of four convergence phases represents a basis for a firm to understand its position within the formation of a new

[2] Cf. Nelson (1995); Nelson and Winter (1982); Nelson *et al.* (1976)

industry cycle. Additionally considering the firm's age and maturity in comparison to the stage of the convergence process (i.e., trajectory reactionist vs. opportunist, definitions 14 and 15), yields a basis for assessing and comparing a firm's strategic positioning within an emerging ecosystem.

Based on the elaborated contingency perspective, managers of firms in converging environments need to be aware of the fact that their firm may have a highly specific positioning within the convergence context. In such situations, doing what competitors do may not necessarily lead to the desired long-term goal. Hence, not only does firm size and maturity matter, i.e., entrant versus incumbent, but so does the entire coevolutionary configuration. Hence, as a basis for strategy formulation and the initiation of capability development processes, generic guidelines for each of the four coevolutionary classes were formulated based on the study of the given sample. These capabilities can either be applied for verifying, refining or formulating strategies for entire firms. On the other hand, particularly for larger firms, these guidelines may serve as a basis for developing portfolio-based assessments and of various activities (e.g., business units, projects, mergers and acquisitions) in the context of convergence. For firms with a rather heterogeneous set of knowledge bases, products or business models, the contingent guidelines may serve as facilitator for managerial decision-making (e.g., through developing roadmaps for distinct projects of capability development based on the coevolutionary framework).

Finally, as the observed firm-level implications and derived managerial capabilities are based on the evolutionary perspective, findings may be applied even for subsets of a convergence process. For instance, in situations where the industry evolution never reaches mature stages, such as applicational or industrial convergence (e.g., due to regulatory constraints), firms may still benefit from convergence-oriented capabilities for creating value between knowledge bases. In other words, even in situations where the development of a convergence phenomenon does not entirely take place, but solely certain indicators of it are satisfied, these findings may serve as a guideline for capability development.

6.2 Outlook

6.2.1 Limitations of the study

Although the theory generation and model development of this work is based on a specific research frame, consisting of the ICT industry sector as well as the geographic scope of Europe and the USA, considerations and recommendations beyond this given frame were made. In particular, a link to emerging trends of convergence around recent developments in nanotechnology is provided, which gives rise to transferring insights from one industry area to another. However, as the propositions elaborated in this study are solely based on observations within the ICT industry, further generalization needs to be handled with care. Although the multiple-case approach aims at increasing the external validity of the findings, this is particularly the case for theoretical replication within areas of highly similar conditions (e.g., other firms within the same industry area). Additionally, as the identified developments around nanotechnologies represent an early phase of the convergence process, the application of insights gained through a retrospective consideration of the convergence process can per definition only yield limited outcomes. Hence, for the case of convergence of nanotechnologies, the verification of such propositions can only be given through future observation over time.

Aiming at establishing a chain of evidence in the analysis of the empirical data, multiple sources were included and triangulated into the study for each firm. Construct validity could however be further increased by interrogating a higher amount of informants within various parts of each organization. Due to the additional investigative lens on industry-level aspects of the convergence phenomenon, it was however chosen to focus on replicating the observations at multiple cases, instead of at multiple parts of a single organization. This allowed a higher degree of saturation and external validity, yet partly at the cost of construct validity.

From a managerial perspective, the resulting model and the identified set of contingent capabilities allows the generic comparison of a given firm with the observations and propositions based on the presented underlying sample. This, in turn, may provide a basis for strategy formulation and the creation of capability development processes. However, as the sample of firms was deliberately selected under the premise of replication logic (i.e., as the firms were claimed to be relevant in terms of convergence beforehand), this study does not present a set of "best practice" stories. With regard to the selection of the firm sample, this study is based on the foundations laid by Rumelt, Schendel, and Teece (1995), who argue that the time is long past when management practice can depend on the method of describing and prescribing the practices of *best* firms. This study therefore represents a descriptive account of how different firms at different coevolutionary constellations act and react under the context of convergence. By connecting attributes of managerial approach and implementation with outcomes on various facets of firm performance, a conceptual empirical link between deliberate management and outcome—either successful or not—within the ecosystem is established.

6.2.2 Further research

Building on the above mentioned limitations, further exploratory and explanatory work on the development of the emerging convergence process around recent trends of nanotechnology is needed. In particular, an extensive elaboration on antecedents and implications in comparison to the case of ICT industry may be of relevance. Whereas recent trends in ICT convergence can be seen as driven through the technological advances as introduced through new products, nanotechnology may not that much be about products, but rather processes instead (e.g., manufacturing in nanoscale).

During earlier phases of the research process, generic patterns with regard to different characteristics of the firm and the convergence phenomenon were searched for, with the aim of resulting in a framework for positioning and clustering the sample. In doing so, compromises had to be made in order to allow simplification into the 2×2 matrix. Not only was the careful discussion between observation and assumption needed at cases where a single firm tended to be located at the boundary of two coevolutionary classes. Furthermore, some cases required a significantly higher amount of cycles between abstraction and data analysis, in order to result in a satisfying classification. For instance, in the case of firm α_3, initial observations yielded the assumption that none of the four coevolutionary classes satisfies the conditions of that particular case. As the business model of the firm, and the resulting strategic positioning in comparison to other firms was perceived as highly different from all other firms in the case set, an extra amount of data analysis and triangulation was needed. Also, at a first glance, the firm was perceived to be located at several stages of the convergence process at the same time. Such exceptions within a given case sample, as represented by firm α_3, may represent a basis for further in-depth qualitative case study development. When a firm does not seem to fit into a given framework, the exceptional characteristics may for instance represent a basis for differentiation and uniqueness within the marketplace—which others might be able to learn from.

Based on the inductive approach of this research process, new theory was generated in the form of a set of propositions. Although these are grounded on the underlying research frame, their investigation with respect to further external validity represents a point of departure for further research. In particular, an elaboration and transformation of the propositions into a set of distinct hypotheses may represent a basis for quantitative and theory-testing research, which would provide validation and verification of the theory throughout a larger sample of study. Obviously, the availability of quantitative data for observations on convergence represents a major challenge. Such a data sample may be created through the structuring and coding of qualitative data, which however is highly time-consuming. Especially recognizing the convergence phenomenon as consisting of a spill-over between previously distinct knowledge bases, quantitative analysis could be based on the examination of the development of two or more given knowledge bases over time. For instance, in the context of convergence between information and telecommunication technology, academic publications from each discipline could be coded and compared over time, e.g., by measuring the rise of common terminology used, or the cross-referencing activity between publications across industry boundaries.

Another basis for developing quantitative studies can be seen in the context of population ecology of firms. Building on the work of Utterback (1994), the peculiarity of the convergence phenomenon may yield interesting results with regard to the amount of firms at mature stages of the evolutionary cycle. With reference to the discussion on the distinctiveness of convergent cycles (section 5.2.2), an issue of interest to the convergence phenomenon may consist in the question whether the scope of consolidation quantitatively exceeds previously distinct industry boundaries. Classical theory on punctuated equilibria and technological change cycles (section 2.3) gives rise to an understanding, that the stabilizing state yields less firms than during the growth peak, but still more firms than at the initial state. In this context, however, the convergence case may be different. As industry boundaries change during the process, consolidating activities may exceed industry boundaries, and eventually yield less firms than in the beginning of the cycle.

By reducing the phenomenon of convergence into such fundamental underlying principles, similar examples for convergent change processes may be looked for in other disciplines. In particular, the convergence process in innovation can be related to the characteristics of similar natural merger processes in a variety of other disciplines.[3] Hence, by comparing the socioeconomic phenomenon of knowledge base-driven convergence with similar phenomena in nature, further research could gain new insights on underlying mechanisms in terms of preconditions and drivers, and simulation models as a support for managerial decision-making could be derived.

Finally, a rather philosophical question for further research may consist in considerations on what happens after convergence, i.e., whether reverse developments can occur or even can be observed. This may particularly be driven by the question on why knowledge bases or scientific disciplines have to come together in the first place, given the fact that they may have same scientific roots back in history. Hence, it remains for history to show whether human behavior intrinsically implies a deliberate search for mainly individual advances and, thereby, divergence in knowledge bases.

> "The laws of nature have always been universal. It is we humans who have compartmentalized knowledge. I have always viewed knowledge as a continuum as far as I can remember." (anonymous post in an Internet newsgroup)

[3] For instance, in chemical reaction processes, similar dynamics can be observed. When observing the trajectory of energy consumption when two or more substances are about to react, a similar declining bell-curve can be observed. In particular, the co-reaction of involved substances requires a certain amount of initial energy, before the reaction can take place, and result in a final energy state which is below initial level. Similarly to the innovation trajectory case, the final energy state turns out to be below initial level, as the originally involved substances free energy as they strive for an energetically better state. As the initial substances merge together, the same total amount of energy is not needed in the system anymore, as it would otherwise become redundant. Hence, the system of the final state can be regarded as consolidated.

Appendix

Table A.1: Overview of sample firms: case ICT

Case company	*Intersectional context*	*Headquarters' location*	*Year of foundation**
δ_6	leveraging new service models based on the interaction between the increasing interaction of personal computers, communication devices, cameras and printers	USA	1939
δ_2	leveraging multiradio-, multifunctional mobile handset devices for consumer, business and entertainment-oriented communication	Europe	1966**
γ_1	achieving synergy for multiple application areas through integrating CPU, wireless units and emerging technologies onto single chipset	USA	1968
γ_2	providing an integrated, comprehensive range of enterprise software applications and business solutions to empower multiple aspects of business operations	Europe	1972
γ_3	offering products that cost-effectively meet the needs of service providers, recognizing critical requirements for data networking technology that enable a new generation of revenue-generating services	USA	1974
δ_3	challenging incumbent business models through combining user-friendly personal computing with music and other entertainment experiences	USA	1976
δ_4	developing processes for facilitating user-centric industrial product design across multiple scientific disciplines	USA	1978
δ_5	providing "the last inch" between human and device through integrating ergonomics, design and quality into computing and entertainment accessory devices	Europe	1981
γ_4	enabling the coming-together of modern computer communications and classic telecommunication based on packet-switched core technologies	USA	1984
γ_5	integrating high-speed wireless and personal electronics into multi-use communication chipsets for mobile devices	USA	1985
δ_1	adopting Internet technologies for providing value-added services to the telecommunications services subscribers	Europe	1990***
δ_7	integrating usability and computing power of disconnected handheld technology with efficient wireless connectivity applications	USA	1992

Table A.1: (continued)

Case company	*Intersectional context*	*Headquarters' location*	*Year of foundation**
β_2	providing an intelligent software link for bridging application usability between the desktop, handheld, wireless and disconnected world of business applications	USA	1993
δ_8	leveraging data transport and switching services of converged all-IP based communication for enterprise customers	USA	1993****
α_1	allowing integration of fragmented and technologically distinct business processes into real-time management applications	USA	1994
β_3	extending categorization capabilities of search-engine into a multitude of communication, entertainment and commerce applications for consumer, small and large businesses	USA	1994
α_2	extending the global online marketplace through integrating payment and communication systems	USA	1995
β_4	enabling converged communication systems through providing solutions to small business markets	USA	1996
δ_9	adopting new technologies for delivering all services and products for mobile, fixed and IP-based voice and data communications to mass market	Europe	1997***
α_3	bridging sectors and communities from the information society through pervasive and intuitive search technologies	USA	1998
β_6	leveraging phone connection to Internet services through integrating speech synthesis and recognition into new applications of interactive voice response	USA	1999
β_5	allowing end-to-end mobile access to e-mail servers through a carrier-hosted solution	USA	2000
α_4	developing terminal software solutions for leveraging streaming multimedia to lightweight mobile handset	USA	2000
α_5	providing multiradio access roaming between cellular carrier networks and broadband hotspots	USA	2001
α_6	providing software for mobile operators and equipment manufacturers to deliver services leveraging multiple broadcast technologies	USA	2003

Table A.1: (continued)

Case company	*Intersectional context*	*Headquarters' location*	*Year of foundation**
β_1	leveraging tools for market segmentation and mass-customization within the intersection of mobile subscription services and Internet technologies	USA	2005

* Sorted by year of foundation.
** As a merger of two firms.
*** After privatization.
**** As a joint venture.

Table A.2: Overview of sample firms: case nanotechnology

Case company	*Intersectional context*	*Headquarters' location*	*Year of foundation**
λ_1	exploring usage of nanoscale technology for drug delivery mechanisms	USA	1946
λ_2	exploring piezoelectronic ink jet printing technologies as a nanoscale manufacturing method	USA	1984
λ_3	guiding nanotechnology research, public policy and education to address the critical challenges facing humanity	USA	1986
λ_4	combinining biotechnology with pharmaceutical applications; nanoscale technology for drug delivery	USA	1987
λ_5	developing nano-bio-electronic detection technologies	USA	2000
λ_6	developing carbon nanotube-based computer displays	USA	2001
λ_7	developing cost-efficient thin-film solar cells	USA	2002
λ_8	exploring usage of nanotechnology for explosives detection sensors	USA	2002
λ_9	applying silicon nanomaterial platform for developing light-activated power-generating products	USA	2002

* Sorted by year of foundation.

Table A.3: List of conducted interviews: case ICT

Case company	*Informant identifier*	*Position level*	*Date of interview**
β_2	1	Senior Vice President	09.02.2006
δ_3	2	Manager	10.02.2006
δ_3	3	Manager	17.02.2006
δ_2	4	Senior Manager	10.02.2006
δ_9	5	Director	14.02.2006
δ_9	5	Director	07.03.2006
δ_4	6	Group Leader	15.02.2006
δ_5	7	Vice President	15.02.2006
β_3	8	Director	24.02.2006
δ_8	9	Vice President and General Manager	06.03.2006
β_6	10	Chief Creative Officer	07.03.2006
α_2	11	President	09.03.2006
β_5	12	Vice President	22.03.2006
α_6	13	Vice President	23.03.2006**
α_6	14	Director	23.03.2006**
β_1	15	Vice President	23.03.2006
δ_6	16	Vice President	27.03.2006
α_5	17	Vice President	28.03.2006
α_4	18	Director	28.03.2006
γ_4	19	Director	29.03.2006
γ_4	20	Vice President and General Manager	02.05.2006
β_4	21	Senior Vice President	29.03.2006
δ_7	41	Director	30.03.2006
γ_2	23	Vice President	30.03.2006
γ_1	24	Director	31.03.2006
γ_1	25	Director and General Manager	29.05.2006
γ_3	26	Chief Technology Officer	31.03.2006
α_3	27	Senior Manager	17.04.2006
γ_5	28	Director	21.04.2006
δ_1	29	President	25.04.2006
α_1	30	Manager	04.05.2006

* Sorted by date of first interview with respective firm.
** Informants 13 and 14 were both interviewed at one single occasion.

Table A.4: List of conducted interviews: case nanotechnology

Case company	*Informant identifier*	*Position level*	*Date of interview**
λ_1	31	Vice President	08.03.2006
λ_7	32	Vice President	20.03.2006
λ_8	33	President & Chief Executive Officer	21.03.2006
*n/a***	34	Professor	22.03.2006
*n/a***	35	Professor emeritus	27.03.2006
λ_4	36	Manager	29.03.2006
λ_6	37	Vice President	10.04.2006
λ_9	38	Chief Technology Officer and Vice President	12.04.2006
λ_2	39	Chief Technology Officer	12.04.2006
λ_5	40	Vice President	13.04.2006
*n/a***	41	Executive Director	21.04.2006
*n/a***	42	Managing Director	24.04.2006
*n/a***	43	Special Assistant	03.05.2006
λ_3	44	President	03.05.2006

* Sorted by date of interview.
** Expert interviews, no affiliation to firms (mostly academic affiliations).

Table A.5: Stages of convergent change: case ICT

*Case company**	*Stage of convergence*				*Scope of trajectories*	*Domain legacy*	*Observed manifestation*
	Knowledge	*Technological*	*Applicational*	*Industrial*			
δ_6			•		computing; wireless communication; cameras; printers	printers; computing	showing the market how business can be made out of the interplay
δ_2				•	computing; wireless communications; entertainment	wireless communications	impact of service horizon increasingly crystallized; industry-external competitors emerge
γ_1		•			computing; wireless communications; entertainment	computing	exploring new CPU-based technological combinations for spurring new applications
γ_2		•			business process technology; desktop computing applications	business process technology	exploring new technological combinations for spurring new applications
γ_3		•			circuit-switched wired communication; packet-switched wired communication	packet-switched wired communication	implementing network interoperability
δ_3				•	computing; wireless communications; entertainment	computing	impact of service horizon increasingly crystallized; industry-external competitors emerge

Table A.5: (continued)

*Case company**	*K*	*T*	*A*	*I*	*Scope of trajectories*	*Domain legacy*	*Observed manifestation*
δ_4				•	computing; wireless communications; entertainment; user centric aspects	user centric aspects	less focus on technology than on broadening perspective of cross-industry user scenarios
δ_5				•	computing; wireless communications; entertainment; user centric aspects	user centric aspects	enhancing end user experience in established applicational intersections
γ_4		•			circuit-switched wired communication; packet-switched wired communication	packet-switched wired communication	exploring new network technologies for spurring new applications
γ_5		•			computing; wireless communications; entertainment	wireless communications	leveraging tools for allowing the creation of applications
δ_1				•	circuit-switched wired communication; packet-switched wired communication; wireless communication	circuit-switched wired communication	trade-off between launching new services and cannibalizing own business; industry-external competitors emerge
δ_7			•		computing; wireless communications	computing	creating new wireless business application experiences based on an established technological confluence
β_2			•		computing; wireless communications	computing	creating new wireless business application user experiences based on an established technological confluence
δ_8			•		circuit-switched wired communication; packet-switched wired communication	packet-switched wired communication	marketing and selling end-user transparency

Table A.5: (continued)

*Case company**	*K*	*T*	*A*	*I*	*Scope of trajectories*	*Domain legacy*	*Observed manifestation*
α_1		•			business process technology; desktop computing applications	business process technology	building tools for leveraging the technological integration
β_3			•		Internet search; e-commerce; business applications; entertainment	Internet search	exploring new service models and target groups in the intersection around hosted Internet-based solutions
α_2		•			e-commerce; financial services; communication	e-commerce	developing new technological combinations for enhancing e-commerce experience
β_4			•		circuit-switched wired communication; packet-switched wired communication	packet-switched wired communication	developing new service provisioning models
δ_9			•		circuit-switched wired communication; packet-switched wired communication; wireless communication	circuit-switched wired communication	trade-off between launching new services and cannibalizing own business; industry-external competitors emerge
α_3		•			Internet search; e-commerce; business applications; entertainment	Internet search	exploratively launching new products without too predetermined business models
β_6			•		circuit-switched wired communication; packet-switched wired communication; voice recognition	voice recognition	showing new service models for combining pervasiveness of telephony with Internet content

Table A.5: (continued)

*Case company**	*K*	*T*	*A*	*I*	*Scope of trajectories*	*Domain legacy*	*Observed manifestation*
β_5			•		computing; wireless communications	computing	creating new wireless business service models based on an established technological confluence
α_4		•			packet-switched wired communication; wireless communication	—	developing technologies for combining streaming media with wireless communications
α_5		•			packet-switched wired communication; wireless communication	wireless communications	developing technologies for mobile-fixnet convergence
α_6	•				packet-switched wired communication; wireless communication	—	building a common knowledge basis translation into wireless broadcast technologies
β_1				•	packet-switched wired communication; wireless communication	—	exploring and developing new niche markets within converged industry

* Sorted by year of foundation.

References

Abernathy, W. J. and Clark, K. B. (1985). Innovation: mapping the winds of creative destruction. *Research Policy*, **14**(1), 3–22.

Abernathy, W. J. and Utterback, J. M. (1978). Patterns of industrial innovation. *Technology Review*, **80**(7), 40–47.

Accenture (2006). Content flips the channel: The Accenture media content survey 2006. Report, Accenture Convergence Group.

Adner, R. (2002). When are technologies disruptive? A demand-based view of the emergence of competition. *Strategic Management Journal*, **23**(8), 667–688.

Adner, R. and Levinthal, D. A. (2000). Technology speciation and the path of emerging technologies. In: G. S. Day, P. J. H. Schoemaker, and R. E. Gunther (eds.), *Wharton on Managing Emerging Technologies*, pp. 57–74. John Wiley & Sons, New York, NY.

Afuah, A. and Tucci, C. L. (2003). A model of the internet as creative destroyer. *IEEE Transactions on Engineering Management*, **50**(4), 395–402.

Ahuja, G. and Katila, R. (2004). Where do resources come from? The role of idiosyncratic situations. *Strategic Management Journal*, **25**(8-9), 887–907.

Aldrich, H. (1979). *Organizations and Environments*. Prentice Hall, Englewood Cliffs, NJ.

Alkemade, M. (2003). *The cuddle economy: the role of alliances and partnerships in the development of mobile data services*. Master's thesis, Erasmus University.

Amesse, F., Latour, R., Rebolledo, C., and Séguin-Dulude, L. (2004). The telecommunications equipment industry in the 1990s: from alliances to mergers and acquisitions. *Technovation*, **24**(11), 885–897.

Andergassen, R., Nardini, F., and Ricottilli, M. (2003). Innovation waves, self-organised criticality and technological convergence. Report, Università di Bologna, Dipartimento di Scienze Economiche.

Anderson, P. and Tushman, M. L. (1990). Technological discontinuities and dominant designs: a cyclical model of technological change. *Administrative Science Quarterly*, **35**(4), 604–633.

Anderson, P. and Tushman, M. L. (1991). Managing through cycles of technological change. *Research/Technology Management*, **34**(3), 26–31.

Andersson, P. and Molleryd, B. G. (1997). Telecommunication services in context: distribution consequences of technological change and convergence. *International Journal of Service Industry Management*, **8**(5), 453–473.

Andrews, S. B. (1995). Wanted: A good network theory of organization (review symposium on Burt, 1992). *Administrative Science Quarterly*, **40**(2), 355–358.

Antonelli, C. (2001). *The Microeconomics of Technological Systems*. Oxford University Press, Oxford.

Argyris, C. and Schön, D. A. (1978). *Organizational learning: a theory of action perspective*. Addison-Wesley, Reading, MA.

Athreye, S. and Keeble, D. (2000). Technological convergence, globalisation and ownership in the UK computer industry. *Technovation*, **20**(5), 227–245.

Bachmann, G. (1996). Nanotechnology: technology analysis. Report, VDI Technologiezentrum. Original German edition 1994.

Backholm, A. and Hacklin, F. (2002). Estimating the 3G convergence effect on the future role of application-layer mobile middleware solutions: scenarios and strategies for business application providers in 3G and beyond. In: W. W. Lu (ed.), *Proceedings of the 2002 International Conference on Third Generation Wireless and Beyond*, pp. 74–79, San Francisco, CA.

Bacon, F. (1909). Essays, civil and moral. In: C. W. Eliot (ed.), *The Harvard Classics*, vol. 3. P. F. Collier & Son, New York, NY.

Baer, M. (2004). *Kooperationen und Konvergenz*, vol. 26 of *Controlling und Management*. Peter Lang, Frankfurt a. M.

Baldwin, C. Y. and Clark, K. B. (1997). Managing in an age of modularity. *Harvard Business Review*, **75**(5), 84–93.

Baldwin, T. F., McVoy, D. S., and Steinfield, C. (1996). *Convergence: Integrating Media, Information and Communication*. Sage Publications, Thousand Oaks, CA.

Ballon, P. (2004). Scenarios and business models for 4G in Europe. *Info – The journal of policy, regulation and strategy for telecommunications*, **6**(6), 363–382.

Bally, N. (2005a). Deriving managerial implications from technological convergence along the innovation process: a case study on the telecommunications industry. Memo-Stencil (Working Paper Series), Åbo Akademi University, Department of Business Studies.

Bally, N. (2005b). Deriving managerial implications from technological convergence along the innovation process: a case study on the telecommunications industry. Unpublished report, ETH Zurich, Chair for Technology Management and Entrepreneurship.

Bane, P. W., Bradley, S. P., and Collis, D. J. (1997). *The converging worlds of telecommunication, computing and entertainment*. Division of Research, Harvard Business School, Boston, MA.

Bannister, J., Mather, P., and Coope, S. (2004). *Convergence technologies for 3G networks IP, UMTS, EGPRS and ATM*. John Wiley & Sons, New York, NY.

Barnes, S. J. (2002). Under the skin: short-range embedded wireless technology. *International Journal of Information Management*, **22**(3), 165–179.

Barnett, H. (1953). *Innovation, the basis of cultural change*. McGraw-Hill, New York, NY.

Barnett, W. P. and Freeman, J. (2001). Too much of a good thing? Product proliferation and organizational failure. *Organization Science*, **12**(5), 539–558.

Bartunek, J. M. (1984). Changing interpretive schemes and organizational restructuring: the example of a religious order. *Administrative Science Quarterly*, **29**(3), 355–372.

Bauer, J. M., Weijnen, M. P. C., Turk, A. L., and Herder, P. M. (2003). Delineating the scope of convergence in infrastructures: new frontiers for competition. In: W. A. H. Thissen and P. M. Herder (eds.), *Critical Infrastructures: State of the Art in Research and Application*, pp. 209–231. Kluwer, Boston, MA.

Baum, J. A. C. (1996). Organizational ecology. In: S. R. Clegg, C. Hardy, and W. R. Nord (eds.), *Handbook of Organization Studies*, pp. 77–114. Sage Publications, Thousand Oaks, CA.

Baum, J. A. C. and McKelvey, B. (eds.) (1999). *Variations in Organization Science: In Honor of Donald T. Campbell*. Sage Publications, Thousand Oaks, CA.

Benner, M. and Tushman, M. I. (2003). Exploitation, exploration, and process management: the productivity dilemma revisited. *Academy of Management Review*, **28**(2), 238–256.

Betz, F. (1993). *Strategic Technology Management*. McGraw-Hill engineering and technology management series. McGraw-Hill, New York, NY.

Birch, H. (1945). The relation of previous experience to insightful problem-solving. *Journal of Comparative Psychology*, **38**, 367–383.

Blomqvist, K., Hara, V., Koivuniemi, J., and Äijö, T. (2003). Towards networked R&D management: Sonera corporation's R&D management in the dynamic environment as an example. In: J. Butler (ed.), *Proceedings of the R&D Management Conference 2003 (RADMA)*, Proceedings, Manchester, England. Blackwell Publishers.

Bohlin, E. (2000). Convergence in communications and beyond: an introduction. In: E. Bohlin, K. Brodin, A. Lundgren, and B. Thorngren (eds.), *Convergence in Communications and Beyond*. North-Holland, Amsterdam.

Borés, C., Saurina, C., and Torres, R. (2003). Technological convergence: a strategic perspective. *Technovation*, **23**(1), 1–13.

Bourgeois, L. J. and Eisenhardt, K. M. (1988). Strategic decision processes in high velocity environments: four cases in the microcomputer industry. *Management Science*, **34**(7), 816–835.

Bourreau, M. and Gensollen, M. (2004). Trends in information and communication technologies. *Communications & Strategies*, **53**(1), 53–69.

Bower, J. L. (2001). Not all M&As are alike—and that matters. *Harvard Business Review*, **79**(3), 92–101.

Bower, J. L. and Christensen, C. M. (1995). Disruptive technologies: Catching the wave. *Harvard Business Review*, **73**(1), 43–53.

Brand, S. (1987). *The Media Lab: inventing the future at MIT*. Viking, New York, NY.

Brown, S. L. and Eisenhardt, K. M. (1997). The art of continuous change: Linking complexity theory and time-paced evolution in relentlessly shifting organizations. *Administrative Science Quarterly*, **42**(1), 1–34.

Brush, T. H. and Artz, K. W. (1999). Toward a contingent resource-based theory: the impact of information asymmetry on the value of capabilities in veterinary medicine. *Strategic Management Journal*, **20**(3), 223–250.

Brusoni, S. and Pavitt, K. (2003). Problem solving and the co-ordination of innovative activities. Report, University of Sussex. SPRU Electronic Working Paper Series, no. 93.

Burgelman, R. A. (1983). A process model of internal corporate venturing in the diversified major firm. *Administrative Science Quarterly*, **28**(2), 223–244.

Burgelman, R. A. (1994). Fading memories: A process theory of strategic business exit in dynamic environments. *Administrative Science Quarterly*, **39**(1), 24–56.

Burgelman, R. A. (2002). Strategy as vector and the inertia of coevolutionary lock-in. *Administrative Science Quarterly*, **47**(2), 325–357.

Burgelman, R. A. and Grove, A. (1996). Strategic dissonance. *California Management Review*, **38**(2), 8–28.

Burns, T. and Stalker, G. M. (1994). Mechanistic and organic systems. In: T. Burns and G. M. Stalker (eds.), *The Management of Innovation*, pp. 96–125. Tavistock Publications, London, 3rd edition.

Burt, R. S. (1992). *Structural holes: the social structure of competition*. Harvard University Press, Cambridge, MA.

Burt, R. S. (1997). The contingent value of social capital. *Administrative Science Quarterly*, **42**(2), 339–365.

Cacciatori, E. and Jacobides, M. G. (2005). The dynamic limits of specialization: vertical integration reconsidered. *Organization Studies*, **26**(12), 1851–1883.

Cameron, G., Proudman, J., and Redding, S. (2005). Technological convergence, R&D, trade and productivity growth. *European Economic Review*, **49**(3), 775–807.

Cantwell, J. A. and Fai, F. (1999). Firms as the source of innovation and growth: the evolution of technological competence. *Journal of Evolutionary Economics*, **9**(3), 331–366.

Carr, N. (2003). IT doesn't matter. *Harvard Business Review*, **81**(5), 41–49.

Chakravarthy, B. (1993). The global information and communications (infocom) industry: emerging scope. Report, INSEAD, Fontainebleau.

Chakravarthy, B. (1994). Flexible commitments: coping with chaos. Report, INSEAD, Fontainebleau.

Chan-Olmsted, S. M. (1998). Mergers, acquisitions and convergence: the strategic alliances of broadcasting, cable television, and telephone services. *Journal of Media Economics*, **11**(3), 33–46.

Chandler, A. D. (1962). *Strategy and structure: chapters in the history of the American industrial enterprise*. MIT Press, Cambridge, MA.

Chandler, A. D. (1997). The computer industry: the first half-century. In: D. B. Yoffie (ed.), *Competing the age of digital convergence*, pp. 37–122. Harvard Business School Press, Boston, MA.

Chen, M. J. and Hambrick, D. C. (1995). Speed, stealth, and selective attack: How small firms differ from large firms in competitive behavior. *Academy of Management Journal*, **38**(2), 453–482.

Choi, D. and Välikangas, L. (2001). Patterns of strategy innovation. *European Management Journal*, **19**(4), 424–429.

Christensen, C. M. (1997). *The Innovator's Dilemma: When New Technologies Cause Great Firms to Fail.* Harvard Business School Press, Boston, MA.

Christensen, C. M. and Rosenbloom, R. S. (1995). Explaining the attacker's advantage: technological paradigms, organizational dynamics, and the value network. *Research Policy*, **24**(2), 233–257.

Ciancetta, M., Colombo, G., Lavagnolo, R., and Grillo, D. (1999). Convergence trends for fixed and mobile services. *IEEE Personal Communications*, **6**(2), 14–21.

Cockburn, I. M., Henderson, R. M., and Stern, S. (2000). Untangling the origins of competitive advantage. *Strategic Management Journal*, **21**(10-11), 1123–1145.

Cohen, W. and Levinthal, D. A. (1990). Absorptive capacity: A new perspective on learning and innovation. *Administrative Science Quarterly*, **35**(1), 128–152.

Collis, D., Bane, P., and Bradley, S. (1997). Winners and loosers: industry structure in the converging world of telecommunications, computing and entertainment. In: D. B. Yoffie (ed.), *Competing the age of digital convergence*, pp. 159–200. Harvard Business School Press, Boston, MA.

Cooper, A. C. and Schendel, D. (1976). Strategic responses to technological threats. *Business Horizons*, **19**(1), 61–69.

Cusumano, M. A. (2003a). Technology strategy and management: Beware the lure of the horizontal. *Communications of the ACM*, **46**(7), 15–17.

Cusumano, M. A. (2003b). Technology strategy and management: Finding your balance in the products and services debate. *Communications of the ACM*, **46**(3), 15–17.

Cusumano, M. A. and Gawer, A. (2002). The elements of platform leadership. *Sloan Management Review*, **43**(3), 51–58.

Cyert, R. M. and March, J. G. (1963). *A behavioral theory of the firm.* Prentice-Hall, Englewood Cliffs, NJ.

Dahlander, L. and Wallin, M. (2006). A man on the inside: unlocking communities as complementary assets. *Research Policy*, **35**(8), 1243–1259.

Danneels, E. (2004). Disruptive technology reconsidered: A critique and research agenda. *Journal of Product Innovation Management*, **21**(4), 246–258.

D'Aveni, R. A. (1994). *Hypercompetition: Managing the Dynamics of Strategic Maneuvering*. The Free Press, New York, NY.

Day, G. S., Schoemaker, P. J. H., and Gunther, R. E. (2000). A different game. In: G. S. Day, P. J. H. Schoemaker, and R. E. Gunther (eds.), *Wharton on Managing Emerging Technologies*. John Wiley & Sons, New York, NY.

DiMaggio, P. J. and Powell, W. W. (1983). The iron cage revisited: institutional isomorphism and collective rationality in organizational fields. *American Sociological Review*, **48**(2), 147–160.

Dixit, A. and Pindyck, R. (1994). *Investment under uncertainty*. Princeton University Press, Princeton, NJ.

Doolittle, R. F. (1994). Convergent evolution: the need to be explicit. *Trends in Biochemical Sciences*, **19**(1), 15–18.

Dosi, G. (1982). Technological paradigms and technological trajectories: A suggested interpretation of the determinants and directions of technical change. *Research Policy*, **11**(3), 147–162.

Dosi, G. (1988). Sources, procedures, and microeconomic effects of innovation. *Journal of Economic Literature*, **26**(3), 1120–1171.

Dosi, G., Freeman, C., Nelson, R., Silverberg, G., and Soete, L. (eds.) (1988). *Technical Change and Economic Theory*. Pinter Publishers, London.

Dougherty, D. (1992). A practice-centered model of organizational renewal through product innovation. *Strategic Management Journal*, **13**(5), 77–92.

Dougherty, D. and Hardy, C. (1996). Sustained product innovation in large, mature organizations: Overcoming innovation-to-organization problems. *Academy of Management Journal*, **39**(5), 1120–1153.

Dowling, M., Lechner, C., and Thielmann, B. (1998). Convergence: innovation and change of market structures between television and online services. *Electronic Markets Journal*, **8**(4), 31–35.

Drucker, P. (1985). The discipline of innovation. *Harvard Business Review*, **63**(3), 67–72.

Duysters, G. (1996). *The dynamics of technical innovation: The evolution and development of information technology*. Edward Elgar, Cheltenham.

Duysters, G. and de Man, A. P. (2003). Transitory alliances: an instrument for surviving turbulent industries? *R&D Management*, **33**(1), 49–58.

Duysters, G. and Hagedoorn, J. (1998). Technological convergence in the IT industry: the role of strategic technology alliances and technological competencies. *International Journal of the Economics of Business*, **5**(3), 355–368.

Duysters, G. and Hagedoorn, J. (2000). Core competences and company performance in the world-wide computer industry. *Journal of High Technology Management Research*, **11**(1), 75–91.

Dürmüller, C. (2006). Das Potenzial gemeinsamer Plattformen nutzen. *io new management*, **75**(9), 26–30.

Edelmann, J., Koivuniemi, J., Hacklin, F., and Stevens, R. (2006). New perspectives on mobile service development. In: J. Müller and B. Preissl (eds.), *Governance of Communication Networks: Connecting Societies and Markets with IT*, pp. 295–308. Physica (Springer), Heidelberg.

Edwards, J. (1999). Convergence reshapes the networking industry. *Computer*, **32**(5), 14–16.

Edwards, K. and Gordon, T. (1984). *Characterization of innovations introduced in the U.S. market in 1982*. The Futures Group, U.S. Small Business Administration, Washington DC.

Eisenhardt, K. M. (1989a). Building theories from case study research. *Academy of Management Review*, **14**(4), 532–550.

Eisenhardt, K. M. (1989b). Making fast strategic decisions in high-velocity environments. *Academy of Management Journal*, **32**(3), 543–576.

Eisenhardt, K. M. and Graebner, M. E. (2007). Theory building from cases: opportunities and challenges. *Academy of Management Journal*, **50**(1), 25–32.

Eisenhardt, K. M. and Martin, J. A. (2000). Dynamic capabilities: What are they? *Strategic Management Journal*, **21**(10-11), 1105–1121.

Eisenhardt, K. M. and Tabrizi, B. N. (1995). Accelerating adaptive processes: Product innovation in the global computer industry. *Administrative Science Quarterly*, **40**(1), 84–110.

Eldredge, N. and Gould, S. (1972). Puncuated equilibria: an alternative phyletic gradualism. In: T. J. Schopf (ed.), *Models in paleobiology*, pp. 82–115. Freeman, Cooper & Co., San Francisco.

Erat, P. (2004). *Forming Business Customer Communities in a Pharmaceutical Firm: Towards a Framework of Challenges*. Ph.D. thesis, Universität St. Gallen (Diss. Nr. 2845).

Ethiraj, S. K. and Levinthal, D. (2004). Bounded rationality and the search for organizational architecture: an evolutionary perspective on the design of organizations and their evolvability. *Administrative Science Quarterly*, **49**(3), 404–437.

European Commission (1997). Green paper on the convergence of the telecommunications, media and information technology sectors, and the implications for regulation: Towards an information society approach. Report, European Commission.

Fagerberg, J. (2002). A layman's guide to evolutionary economics. In: *Industrial R&D and Innovation Policy Learning: Evolutionary Perspectives and New Methods for Impact Assessment*, Leangkollen, Asker, Norway. Norges Forskningråd.

Fagerberg, J. and Verspagen, B. (2002). Technology-gaps, innovation-diffusion and transformation: an evolutionary interpretation. *Research Policy*, **31**(8-9), 1291–1304.

Fahrni, F. (2001a). Technologie als Triebfeder für Innovationen. In: N. Thom and R. J. Zaugg (eds.), *Excellence durch Personal- und Organizationskompetenz*, pp. 307–318. Haupt, Berlin.

Fahrni, F. (2001b). Wachsen oder sterben? In: F. Fahrni and G. Schuh (eds.), *Technologiemanagement als Treiber nachhaltigen Wachstums*, pp. 5–15. Shaker Verlag, Aachen.

Fai, F. M. and von Tunzelmann, N. (2001). Industry-specific competencies and converging technological systems: evidence from patents. *Structural Change and Economic Dynamics*, **12**(2), 141–170.

Farber, D. and Baran, P. (1977). The convergence of computing and telecommunications systems. *Science*, **195**(4283), 1166–1170.

Fine, C. H. (1998). *Clockspeed: Winning Industry Control in the Age of Temporary Advantage*. Perseus, New York, NY.

Forester, T. (1993). *Silicon samurai: How Japan conquered the world's IT industry*. Blackwell Publishers, Cambridge, MA.

Foster, R. (1986). *Innovation: The Attacker's Advantage*. MacMillan Press, London.

Fransman, M. (2000). Convergence, the internet and multimedia: implications for the evolution of industries and technologies. In: E. Bohlin, K. Brodin, A. Lundgren, and B. Thorngren (eds.), *Convergence in Communications and Beyond*. North-Holland, Amsterdam.

Friedman, J. and Waldman, M. (1992). *Collision or convergence: digital video/audio, computers, and telecommunications*. Society of Motion Picture and Television Engineers, White Plains, NY. Proceedings of the 26th Annual SMPTE Advanced Television and Electronic Imaging Conference, San Francisco, CA.

Fusaro, P. C. (2002). *Energy convergence: the beginning of the multi-commodity market*. John Wiley & Sons, New York, NY.

Gaines, B. R. (1998). The learning curves underlying convergence. *Technological Forecasting and Social Change*, **57**(1-2), 7–34.

Galunic, D. C. and Eisenhardt, K. M. (1996). The evolution of intracorporate domains: Divisional charter losses in high-technology, multidivisional corporations. *Organization Science*, **7**(3), 255–282.

Gambardella, A. and Torrisi, S. (1998). Does technological convergence imply convergence in markets? Evidence from the electronics industry. *Research Policy*, **27**(5), 445–463.

Garcia, R. and Calantone, R. (2002). A critical look at technological innovation typology and innovativeness terminology: a literature review. *Journal of Product Innovation Management*, **19**(2), 110–132.

Gawer, A. (2000). *The organization of platform leadership: an empirical investigation of Intel's management processes aimed at fostering complementary innovation by third parties*. Ph.D. thesis, Massachusetts Institute of Technology.

Gawer, A. and Cusumano, M. A. (2002). *Platform Leadership: How Intel, Microsoft, and Cisco Drive Industry Innovation*. Harvard Business School Press, Boston, MA.

Georghiou, L., Metcalfe, J., Gibbons, M., Ray, T., and Evans, J. (1986). *Post-innovation performance: technological development and competition*. Macmillan, New York, NY.

Gersick, C. J. G. (1991). Revolutionary change theories: a multilevel exploration of the punctuated equilibrium paradigm. *Academy of Management Review*, **16**(1), 10–36.

Glaser, B. G. and Strauss, A. L. (1967). *The Discovery of Grounded Theory: Strategies for Qualitative Research*. Aldine Publishing Company, Chicago.

Gomes-Casseres, B. and Leonard-Barton, D. (1997). Alliance clusters in multimedia: safety net or entanglement. In: D. B. Yoffie (ed.), *Competing the age of digital convergence*, pp. 325–370. Harvard Business School Press, Boston, MA.

Gong, J. and Srinagesh, P. (1996). Network competition and industry structure. *Industrial and Corporate Change*, **5**(4), 1231–1241.

Granstrand, O. (1998). Towards a theory of the technology-based firm. *Research Policy*, **27**(5), 465–489.

Grant, R. M. (1996). Prospering in dynamically-competitive environments: Organizational capability as knowledge integration. *Organization Science*, **7**(4), 375–387.

Grant, R. M. (2002). *Contemporary strategy analysis: concepts, techniques, applications*. Blackwell Publishers, Malden, MA, 4th edition.

Greenstein, S. M. and Khanna, T. (1997). What does industry convergence mean? In: D. B. Yoffie (ed.), *Competing the age of digital convergence*, pp. 201–226. Harvard Business School Press, Boston, MA.

Greenwood, R. and Hinings, C. R. (1996). Understanding radical organizational change: bringing together the old and the new institutionalism. *Academy of Management Review*, **21**(4), 1022–1054.

Grossman, J. B. (1970). Supreme court and social change: a preliminary inquiry. *American Behavioral Scientist*, **13**(4), 535–551.

Grove, A. (1997). Navigating strategic inflection points. *Business Strategy Review*, **8**(3), 11–18.

Grover, V. and Saeed, K. (2003). The telecommunication industry revisited: the changing pattern of partnerships. *Communications of the ACM*, **46**(7), 119–125.

Guilhon, B. (2001). The emergence of quasi-market for knowledge. In: B. Guilhon (ed.), *Technology and Markets for Knowledge: Knowledge Creation, Diffusion and Exchange within a Growing Economy*, pp. 21–40. Springer.

Gulati, R., Nohria, N., and Zaheer, A. (2000). Strategic networks. *Strategic Management Journal*, **21**(3), 203–215.

Götte, S. (2003). *Konvergenz der TIME-Märkte: empiriegestützte Entwicklung eines adaptiven und innovativen Konzeptrahmens für ein kooperatives und integrierendes, geschäftskundenbezogenes Marketingmanagement Schweizer Anbieter*. Rainer Hampp Verlag, München.

Hackler, K. and Jopling, E. (2003). Technology convergence driving business model collision. Report, Gartner Group.

Hacklin, F. and Marxt, C. (2003). Assessing R&D management strategies for wireless applications in a converging environment. In: J. Butler (ed.), *Proceedings of the R&D Management Conference 2003 (RADMA)*, Manchester, England. Blackwell Publishers.

Hacklin, F., Raurich, V., and Marxt, C. (2004a). How incremental innovation becomes disruptive: the case of technology convergence. In: M. Xie, T. S. Durrani, and H. K. Tang (eds.), *Proceedings of the IEEE International Engineering Management Conference*, vol. 1, pp. 32–36, Singapore. IEEE Engineering Management Society.

Hacklin, F., Lopperi, K., Bergman, J.-P., and Marxt, C. (2004b). Toward an integrated knowledge management cycle in cumulative open innovation networks. In: *Proceedings of The R&D Management Conference 2004 (RADMA)*, Sesimbra, Portugal.

Hacklin, F., Adamsson, N., Marxt, C., and Norell, M. (2005a). Design for convergence: managing technological partnerships and competencies across and within industries. In: *15th International Conference on Engineering Design (ICED)*, Melbourne. The Design Society.

Hacklin, F., Raurich, V., and Marxt, C. (2005b). Implications of technological convergence on innovation trajectories: the case of ICT industry. *International Journal of Innovation and Technology Management*, **2**(3), 313–330.

Hacklin, F., Bergman, J. P., Nyström, A.-G., Marxt, C., and Jantunen, A. (2005c). Knowledge convergence through interdisciplinary capabilities: fostering disruptive innovation in deconstructed value creation environments. In: *Proceedings of the PICMET'05 Conference*, Portland, OR. Portland International Center for Management of Engineering and Technology (PICMET).

Hacklin, F., Marxt, C., and Inganäs, M. (2005d). Open-up or shakeout: managing technological convergence through value networks. In: *Proceedings of the PICMET'05 Conference*, Portland, OR. Portland International Center for Management of Engineering and Technology (PICMET).

Hamel, G. and Prahalad, C. K. (1994). *Competing for the future*. Harvard Business School Press, Boston, MA.

Hannan, M. T. and Freeman, J. (1984). Structural inertia and organizational change. *American Sociological Review*, **49**(2), 149–164.

Hanssen-Bauer, J. and Snow, C. (1996). Responding to hypercompetition: The structure and processes of a regional learning network organization. *Organization Science*, **7**(4), 413–427.

Harianto, F. and Pennings, J. M. (1994). Technological convergence and scope of organizational innovation. *Research Policy*, **23**(3), 293–304.

Harrison, F. and Hearnden, S. (1999). The challenge to realise convergence of fixed and mobile communications. *Electronics & Communication Engineering Journal*, **11**(3), 164–168.

Hawkins, R. (1999). The rise of consortia in the information and communication technology industries: emerging implications for policy. *Telecommunications Policy*, **23**(2), 159–173.

Helfat, C. E. (1997). Know-how and asset complementarity and dynamic capability accumulation: The case of R&D. *Strategic Management Journal*, **18**(5), 339–360.

Helfat, C. E. and Peteraf, M. A. (2003). The dynamic resource-based view: capability lifecycles. *Strategic Management Journal*, **24**(10), 997–1010.

Henderson, R. M. and Clark, K. B. (1990). Architectural innovation: The reconfiguration of existing product technologies and the failure of established firms. *Administrative Science Quarterly*, **35**(1), 9–30.

Henten, A. (2004). Technological convergence and competition. In: *Telecompetition: towards a single Nordic market for telecommunication services?*, pp. 90–104. Föreningen Norden, Nordisk Minsterråd, Copenhagen.

Heraud, J. A. (2003). Regional innovation systems and European research policy: Convergence or misunderstanding? *European Planning Studies*, **11**(1), 41–56.

Hodgson, G. M. and Knudsen, T. (2004). The firm as an interactor: firms as vehicles for habits and routines. *Journal of Evolutionary Economics*, **14**(3), 281–307.

Hofer, C. W. (1975). Toward a contingency theory of business strategy. *The Academy of Management Journal*, **18**(4), 784–810.

Hoogervorst, J. (2004). Enterprise architecture: enabling integration, agility and change. *International Journal of Cooperative Information Systems*, **13**(3), 213–233.

Hughes, A. and Morton, M. S. S. (2006). The transforming power of complementary assets. *Sloan Management Review*, **47**(4), 50–58.

Husig, S., Hipp, C., and Dowling, M. (2005). Analysing disruptive potential: the case of wireless local area network and mobile communications network companies. *R&D Management*, **35**(1), 17–35.

Jacobides, M. G. (2005). Industry change through vertical disintegration: how and why markets emerged in mortgage banking. *Academy of Management Journal*, **48**(3), 465–498.

Jansen, J. J. P. (2005). *Ambidextrous Organizations: A Multiple-Level Study of Absorptive Capacity, Exploratory and Exploitative Innovation and Performance*. Ph.D. thesis, Erasmus University.

Johansson, F. (2004). *The Medici Effect: Breakthrough Insights at the Intersection of Ideas, Concepts, and Cultures*. Harvard Business School Press, Boston, MA.

Jonkers, H., Lankhorst, M. M., ter Doest, H. W. L., Arbab, F., Bosma, H., and Wieringa, R. J. (2006). Enterprise architecture: Management tool and blueprint for the organisation. *Information Systems Frontiers*, **8**(2), 63–66.

Joseph, R. A. (1993). Converging telecommunications technologies: Challenges facing government and regulators in Australia and New-Zealand. *Telecommunications Policy*, **17**(7), 493–503.

Kaluza, B., Blecker, T., and Bischof, C. (1998). Strategic management in converging industries. Report, University of Klagenfurt.

Kaluza, B., Blecker, T., and Bischof, C. (1999). Implications of digital convergence on strategic management. In: S. B. Dahiya (ed.), *The Current State of Economic Science*, pp. 2223–2249. Spellbound Publications, Rohtak.

Kapoor, A. (2004). Telecommunications policy, industrial dynamics and structural transformation: the indian experience. In: *DRUID Summer Conference on Industrial Dynamics, Innovation and Development*, Helsingør, Denmark. Danish Research Unit for Industrial Dynamics (DRUID).

Kassicieh, S. K., Walsh, S. T., Cummings, J. C., McWhorter, P. J., Romig, A. D., and Williams, W. D. (2002a). Factors differentiating the commercialization of disruptive and sustaining technologies. *IEEE Transactions on Engineering Management*, **49**(4), 375–387.

Kassicieh, S. K., Kirchhoff, B. A., Walsh, S. T., and McWhorter, P. J. (2002b). The role of small firms in the transfer of disruptive technologies. *Technovation*, **22**(11), 667–674.

Katila, R. and Ahuja, G. (2002). Something old, something new: A longitudinal study of search behavior and new product introduction. *Academy of Management Journal*, **45**(6), 1183–1194.

Katz, M. L. (1996). Remarks on the economic implications of convergence. *Industrial and Corporate Change*, **5**(4), 1079–1096.

Kauffman, S. A. (1993). *The Origins of Order: Self-Organization and Selection in Evolution*. Oxford University Press, New York, NY.

Kawashima, M. (2002). *Telecom value chain dynamics and carriers' strategies in converged networks*. Master's thesis, Massachusetts Institute of Technology.

Kim, P. (1999). A story of failed technology: Deconstructing interactive TV networks. *Javnost-the Public*, **6**(3), 87–99.

King, A. A. and Tucci, C. L. (1999). Can old disk-drive companies learn new tricks? In: *Proceedings of the 6th Annual International Product Development Management Conference*, Cambridge, UK.

King, A. A. and Tucci, C. L. (2002). Incumbent entry into new market niches: the role of experience and managerial choice in the creation of dynamic capabilities. *Management Science*, **48**(2), 171–186.

Kirchhoff, B. A., Kassicieh, S. K., and Walsh, S. T. (2002). Introduction to the special cluster on the commercialization of disruptive technologies and discontinuous innovations. *IEEE Transactions on Engineering Management*, **49**(4), 319–321.

Klein, K. J., Tosi, H. L., and Cannella, A. A. (1999). Multilevel theory building: Benefits, barriers, and new developments. *Academy of Management Review*, **24**(2), 243–248.

Kleinberg, J. (2005). Economic prospects in nanobiotechnology. In: *NanoBioConvergence Forum Meeting*, Palo Alto, CA.

Klepper, S. (1997). Industry life cycles. *Industrial and Corporate Change*, **6**(1), 145–182.

Klepper, S. and Graddy, E. (1990). The evolution of new industries and the determinants of market-structure. *Rand Journal of Economics*, **21**(1), 27–44.

Kodama, F. (1992). Technology fusion and the new R&D. *Harvard Business Review*, **70**(4), 70–78.

Kodama, F. (1995). *Emerging patterns of inovation*. Harvard Business School Press, Boston, MA.

Kostoff, R. N., Boylan, R., and Simons, G. R. (2004). Disruptive technology roadmaps. *Technological Forecasting and Social Change*, **71**(1-2), 141–159.

Koza, M. P. and Lewin, A. Y. (1998). The co-evolution of strategic alliances. *Organization Science*, **9**(3), 255–264.

Krackhardt, D. (1995). Wanted: A good network theory of organization (review symposium on Burt, 1992). *Administrative Science Quarterly*, **40**(2), 350–354.

Kubicek, H. (1975). *Empirische Organisationsforschung: Konzeption und Methodik*. Poeschel Verlag, Stuttgart.

Lang, G. (2003). TIME-Konvergenz: Einige Überlegungen aus volkswirtschaftlicher Sicht. Report, Universität Augsburg, Institut für Volkswirtschaftslehre. Working paper Nr. 234.

Lant, T. K. and Mezias, S. J. (1992). An organizational model of convergence and reorientation. *Organization Science*, **3**(1), 47–71.

Lavie, D. and Rosenkopf, L. (2006). Balancing exploration and exploitation in alliance formation. *Academy of Management Journal*, **49**(4), 797–818.

Lawless, M. W., Bergh, D. D., and Wilsted, W. D. (1989). Performance variations among strategic group members: An examination of individual firm capability. *Journal of Management*, **15**(4), 649–661.

Lawrence, P. R. and Lorsch, J. W. (1967). *Organization and environment: managing differentiation and integration*. Harvard University, Graduate School of Business Administration, Division of Research, Boston, MA.

Lee, G. K.-F. (2003). *The Competitive Consequences of Technological Convergence in an Era of Innovations: Telephony Communications and Computer Networking, 1989-2001*. Ph.D. thesis, Haas School of Business, UC Berkeley.

Lehner, F. (2000). *Organizational memory*. Hanser Verlag, München.

Lei, D. T. (2000). Industry evolution and competence development: the imperatives of technological convergence. *International Journal of Technology Management*, **19**(7-8), 699–738.

Lei, D. T. and Slocum, J. W. (1991). Global strategic alliances: payoffs and pitfalls. *Organizational Dynamics*, **19**(3), 44–62.

Lei, D. T. and Slocum, J. W. (1992). Global strategy, competence-building and strategic alliances. *California Management Review*, **35**(1), 81–97.

Lemola, T. (2002). Convergence of national science and technology policies: the case of Finland. *Research Policy*, **31**(8-9), 1481–1490.

Leonard-Barton, D. (1992). Core capabilities and core rigidities: a paradox in managing new product development. *Strategic Management Journal*, **13**(SI), 111–125.

Lettl, C., Rost, K., and von Wartburg, I. (2006). Technological merit of user inventions in surgery. In: *Presentation at SMI brownbag seminar, ETH Zurich, June 29, 2006*.

Levinthal, D. A. and March, J. G. (1993). The myopia of learning. *Strategic Management Journal*, **14**(4), 95–112.

Li, F. and Whalley, J. (2002). Deconstruction of the telecommunications industry: from value chains to value networks. *Telecommunications Policy*, **26**(9-10), 451–472.

Lichtenthaler, E. (2004). Technological change and the technology intelligence process: a case study. *Journal of Engineering and Technology Management*, **21**(4), 331–348.

Lind, J. (2004). Convergence: history of term usage and lessons for firm strategists. In: *ITS 15th Biennial Conference*, Berlin, Germany. International Telecommunications Society (ITS).

Lind, J. (2005). Ubiquitous convergence: market redefinitions generated by technological change and the industry life cycle. In: *Proceedings of the DRUID Academy Winter 2005 Conference*, Skørping, Denmark.

Lind, M. R. and Zmud, R. W. (1991). The influence of a convergence in understanding between technology providers and users on information technology innovativeness. *Organization Science*, **2**(2), 195–217.

Lindmark, S., Andersson, E., Bohlin, E., and Johansson, M. (2004). Innovation system dynamics in the Swedish telecom sector. In: *ITS 15th Biennial Conference*, Berlin, Germany. International Telecommunications Society (ITS).

Linton, J. D. (2002). Forecasting the market diffusion of disruptive and discontinuous innovation. *IEEE Transactions on Engineering Management*, **49**(4), 365–374.

Longstaff, P. H. (2001). New ways to think about the visions called "convergence": A guide for business and public policy. Report, Harvard University, Center for Information Policy Research, Boston, MA.

Luthans, F. and Stewart, T. I. (1977). A general contingency theory of management. *Academy of Management Review*, **2**(2), 181–195.

Maidique, M. A. and Zirger, B. J. (1984). A study of success and failure in product innovation: the case of the United States electronics industry. *IEEE Transactions on Engineering Management*, **31**(4), 192–203.

March, J. G. (1991). Exploration and exploitation in organizational learning. *Organization Science*, **2**(1), 71–87.

March, J. G. (1994). *A primer on decision making: how decisions happen*. The Free Press, New York, NY.

March, J. G. (1996). Continuity and change in theories of organizational action. *Administrative Science Quarterly*, **41**(2), 278–287.

Markides, C. C. and Williamson, P. J. (1996). Corporate diversification and organizational structure: A resource-based view. *Academy of Management Journal*, **39**(2), 340–367.

Marquis, D. (1969). *Successful industrial innovations: A study of factors underlying innovation in selected firms*. National Science Foundation, Washington.

Martin, J. (1978). *The wired society*. Prentice-Hall, Englewood Cliffs, NJ.

Marxt, C. and Hacklin, F. (2004). Wenn man zusammenspannen muss: Kooperationen verlangen ähnliche Unternehmenskulturen. *Neue Zürcher Zeitung*, **225**(274), B9. 23.11.2004.

Marxt, C. and Hacklin, F. (2005). Design, product development, innovation: all the same in the end? A short discussion on terminology. *Journal of Engineering Design*, **16**(4), 413–421.

Mazza, G. (2003). Erstellung und Abschätzung von strategischen Szenarien für 'Virtual Identity Providers'. Unpublished report, ETH Zurich, Chair for Technology Management and Entrepreneurship.

McKelvey, B. (1999). Avoiding complexity catastrophe in coevolutionary pockets: strategies for rugged landscapes. *Organization Science*, **10**(3), 294–321.

McKelvey, B. (2002). Managing coevolutionary dynamics. In: *18th EGOS Conference*, Barcelona, Spain.

Messerschmitt, D. G. (1996a). The convergence of telecommunications and computing: What are the implications today? *Proceedings of the IEEE*, **84**(8), 1167–1186.

Messerschmitt, D. G. (1996b). Convergence of telecommunications with computing. *Technology in Society*, **18**(3), 285–296.

Meyer, C. and Davis, S. M. (2003). *It's alive: the coming convergence of information, biology, and business*. Crown Business, New York, NY.

Meyers, P. W. and Tucker, F. G. (1989). Defining roles for logistics during routine and radical technological innovation. *Journal of the Academy of Marketing Science*, **17**(1), 73–82.

Miles, M. B. and Huberman, A. M. (1984). *Qualitative data analysis*. Sage Publications, Thousand Oaks, CA.

Moore, G. A. (1991). *Crossing the Chasm*. Harper Business, New York, NY.

Moore, J. (1993). Predators and prey: a new ecology of competition. *Harvard Business Review*, **71**(3), 75–86.

Moschella, D. C. (1997). *Waves of power: dynamics of global technology leadership 1964-2010*. Amer Management Association, New York, NY.

Mueller, M. (1999). Digital convergence and its consequences. *Javnost—The Public*, **6**(3), 11–27.

Munoz, M. and Rubio, C. G. (2004). A new model for service and application convergence in 3G/4G networks. *IEEE Wireless Communications*, **11**(5), 6–12.

Möller, K. and Svahn, S. (2003). Managing strategic nets: A capability perspective. *Marketing Theory*, **3**(2), 201–226.

Möller, K., Rajala, A., and Svahn, S. (2005). Strategic business nets—their type and management. *Journal of Business Research*, **58**(9), 1274–1284.

Nadler, D. A., Gerstein, M. S., and Shaw, R. B. (1992). *Organizational Architecture: Designs for Changing Organizations*. Jossey-Bass Business and Management Series. Jossey-Bass, San Francisco.

NEC (1984). The first 80 years. Report, Nippon Electronics Company (NEC corporation), Tōkyō, Japan.

Nelson, R. (1995). Recent evolutionary theorizing about economic change. *Journal of Economic Literature*, **33**(1), 48–90.

Nelson, R. and Winter, S. (1982). *An Evolutionary Theory of Economic Change*. Harvard University Press, Cambridge, MA.

Nelson, R. R., Winter, S. G., and Schuette, H. L. (1976). Technical change in an evolutionary model. *Quarterly Journal of Economics*, **90**(1), 90–118.

Niemack, A. and Weber, F. (2005). Konvergenz: Chancen und Risiken. *Comtec*, **2**, 6–8. Swisscom AG, Bern.

Nikolaou, N. A., Vaxevanakis, K. G., Maniatis, S. I., Venieris, I. S., and Zervos, N. A. (2002). Wireless convergence architecture: A case study using GSM and wireless LAN. *Mobile Networks & Applications*, **7**(4), 259–267.

Nokia (2004). Digital convergence: a new chapter for mobility. Report, Nokia Group.

Nonaka, I. (1991). The knowledge-creating company. *Harvard Business Review*, **69**(6), 96–104.

Nonaka, I. (1994). A dynamic theory of organizational knowledge creation. *Organization Science*, **5**(1), 14–37.

Normann, R. (1971). Organizational innovativeness: Product variation and reorientation. *Administrative Science Quarterly*, **16**(2), 203–215.

Nyström, A.-G. (2004). A network description on the finnish mobile telecommunications sector. Report, Åbo Akademi, Department of Business Studies. Memo-Stencil (Working Paper Series), Åbo Akademi University, Department of Business Studies.

Nyström, A.-G. and Hacklin, F. (2005). Operator value creation through technological convergence: the case of VoIP. In: *16th European Regional Conference*, Porto, Portugal. International Telecommunications Society (ITS).

OECD (1992). Telecommunications and broadcasting: convergence or collision? Report, Organisation for Economic Co-operation and Development (OECD).

OECD (1996). Convergence between communications technologies: case studies from North America and Western Europe. Report, Organisation for Economic Co-operation and Development (OECD).

Oettinger, A., Berman, P., and Read, W. (1977). *High and low politics: information resources for the 80s*. Ballinger, Cambridge, MA.

O'Reilly, C. A. and Tushman, M. L. (2004). The ambidextrous organization. *Harvard Business Review*, **82**(4), 74–81.

Paap, J. and Katz, R. (2004). Anticipating disruptive innovation. *Research-Technology Management*, **47**(5), 13–22.

Parodi, R. and Liggieri, G. (2003). Opportunities and constraints for broadband wireless development in the Italian market: A mobile operator perspective. *Journal of the Communications Network*, **2**(3), 89–93.

Pavitt, K. (1980). *Technical innovation and British economic performance*. MacMillan Press, London.

Pavitt, K. (2002). Systems integrators as "post-industrial" firms? In: *DRUID Summer Conference on Industrial Dynamics of the New and Old Economy*, Copenhagen, Denmark. Danish Research Unit for Industrial Dynamics (DRUID).

Pavitt, K. (2004). Innovation processes. In: J. Fagerberg, D. C. Mowery, and R. R. Nelson (eds.), *The Oxford Handbook of Innovation*, pp. 86–114. Oxford University Press, Oxford.

Pennings, J. M. and Puranam, P. (2001). Market convergence and firm strategy: new directions for theory and research. In: *ECIS Conference, The Future of Innovation Studies*, Eindhoven, The Netherlands.

Petrina, S., Volk, K., and Kim, S. (2004). Technology and rights. *International Journal of Technology and Design Education*, **14**(3), 181–204.

Pettigrew, A. M. (1997). What is a processual analysis? *Scandinavian Journal of Management*, **13**(4), 337–348.

Porter, M. E. (1980). *Competitive strategy: techniques for analyzing industries and competitors*. The Free Press, New York, NY.

Porter, M. E. (1985). *Competitive Advantage: Creating and Sustaining Superior Performance*. The Free Press, New York, NY.

Porter, M. E. (1990). *Competitive Advantage of Nations*. The Free Press, New York, NY.

Porter, M. E. (1991). Towards a dynamic theory of strategy. *Strategic Management Journal*, **12**(SI), 95–117.

Prahalad, C. K. (1998). Managing discontinuities: The emerging challenges. *Research-Technology Management*, **41**(3), 14–22.

Prahalad, C. K. and Ramaswamy, V. (2000). Co-opting customer competence. *Harvard Business Review*, **78**(1), 79–87.

Quinn, S. (2005). Convergence's fundamental question. *Journalism Studies*, **6**(1), 29–38.

Ralph, D. and Graham, P. (2004). *MMS: Technologies, Usage and Business Models*. John Wiley & Sons, New York, NY. pp. 18-20.

Ramos, S., Feijoo, C., Perez, J., Castejon, L., and Segura, I. (2002). Mobile Internet evolution models: implications on European mobile operators. *Journal of the Communications Network*, **1**(2), 171–176.

Rao, P., Vemuri, V. K., and Galvin, P. (2004). The changing technological profile of the leading ICT firms: Evidence from US patent data, 1981-2000. *Industry and Innovation*, **11**(4), 353–372.

Rao, P. M. (1999). Convergence and unbundling of corporate R&D in telecommunications: is software taking the helm? *Telecommunications Policy*, **23**(1), 83–93.

Raymond, E. S. (1999). The cathedral and the bazaar. *Knowledge, Technology, and Policy*, **12**(3), 23–49.

Raymond, E. S. (2001). *The Cathedral and the Bazaar: Musings on Linux and Open Source by an Accidental Revolutionary*. O'Reilly, Sebastopol, CA.

Rice, M. P., O'Connor, G. C., Peters, L. S., and Morone, J. G. (1998). Managing discontinuous innovation. *Research-Technology Management*, **41**(3), 52–58.

Rihinen, J. (2006). Arkkitehtuuri tukee yrityksen hallintaa. *Talouselämä*, **69**(28), 30–33.

Robins, F. (2003). The marketing of 3G. *Marketing Intelligence & Planning*, **21**(6), 370–378.

Rockenhäuser, J. (1999). *Digitale Konvergenz und Kompetenzenmanagement*. Deutscher Univ.-Verlag, Wiesbaden.

Roco, M. C. and Bainbridge, W. S. (2002a). Converging technologies for improving human performance: Integrating from the nanoscale. *Journal of Nanoparticle Research*, **4**(4), 281–295.

Roco, M. C. and Bainbridge, W. S. (eds.) (2002b). *Converging technologies for improving human performance: nanotechnology, biotechnology, information technology and cognitive science*. National Science Foundation, Arlington, VA.

Roco, M. C. and Bainbridge, W. S. (2005). Societal implications of nanoscience and nanotechnology: Maximizing human benefit. *Journal of Nanoparticle Research*, **7**(1), 1–13.

Roco, M. C. and Montemagno, C. D. (eds.) (2004). *The Coevolution of Human Potential and Converging Technologies*, vol. 1013 of *Annals of the New York Academy of Sciences*. New York Academy of Sciences, New York, NY.

Rojas-Chapana, J. A. and Giersig, M. (2006). Multi-walled carbon nanotubes and metallic nanoparticles and their application in biomedicine. *Journal of Nanoscience and Nanotechnology*, **6**(2), 316–321.

Rolland, A. (2003). Convergence as strategy for value creation. *International Journal on Media Management*, **5**(1), 14–24.

Rosenberg, N. (1963). Technological change in the machine-tool industry, 1840-1910. *Journal of Economic History*, **23**(4), 414–443.

Rosenberg, N. (1976). *Perspectives on technology*. Cambridge University Press, Cambridge, MA.

Rosenkopf, L. and Tushman, M. L. (1994). The coevolution of technology and organization. In: J. Baum and J. Singh (eds.), *Evolutionary Dynamics of Organizations*, pp. 403–424. Oxford University Press, New York, NY.

Rosenkopf, L., Metiu, A., and George, V. P. (2001). From the bottom up? Technical committee activity and alliance formation. *Administrative Science Quarterly*, **46**(4), 748–772.

Rothaermel (2001). Incumbent's advantage through exploiting complementary assets via interfirm cooperation. *Strategic Management Journal*, **22**(6-7), 687–699.

Rothwell, R. and Gardiner, P. (1988). Reinnovation and robust designs: producer and user benefits. *Journal of Marketing Management*, **3**(3), 372–387.

Rumelt, R. P., Schendel, D., and Teece, D. J. (1995). *Fundamental issues in strategy: a research agenda*. Harvard Business School Press, Boston, MA.

Sabat, H. K. (2002). The evolving mobile wireless value chain and market structure. *Telecommunications Policy*, **26**(9-10), 505–535.

Sahal, D. (1985). Technological guideposts and innovation avenues. *Research Policy*, **14**(2), 61–82.

Salancik, G. R. (1995). Wanted: A good network theory of organization (review article on Burt, 1992). *Administrative Science Quarterly*, **40**(2), 345–349.

Saracco, R. (2005). Leveraging technology and the market to bring telecommunications business into the next decade. *IEEE Communications Magazine*, **43**(2), 1–4.

Saviotti, P. (1996). *Technological Evolution, Variety and the Economy*. Edward Elgar, Cheltenham, UK.

Schmidt, J. B. and Calantone, R. J. (1998). Are really new product development projects harder to shut down? *Journal of Product Innovation Management*, **15**(2), 111–123.

Schmookler, J. (1966). *Invention and economic growth*. Harvard University Press, Cambridge, MA.

Schneider, B. (2002). Network architectures: conversion strategies and evolution towards the next-generation network. *The Journal of the Communications Network*, **1**(2), 47–51.

Schumpeter, J. (1912). *Theorie der wirtschaftlichen Entwicklung*. Duncker & Humblot, Leipzig.

Schumpeter, J. A. (1939). *Business cycles: a theoretical, historical and statistical analysis of the capitalist process*. McGraw-Hill, New York, NY.

Shepard, S. (2002). *Telecommunications convergence: how to bridge the gap between technologies and services*. McGraw-Hill, New York, NY.

Sherif, M. H. (1998). Convergence: A new perspective for standards. *IEEE Communications Magazine*, **36**(1), 110–111.

Sigurdson, J. and Ericsson, P. (2003). New services in 3G: new business models for streaming and video. *International Journal of Mobile Communications*, **1**(1/2), 15–34.

Simon, H. A. (1987). Bounded rationality. In: J. Eatwell, M. Millgate, and P. Newman (eds.), *The New Palgrave: A Dictionary of Economics*. Macmillan, London.

Smith, R. (2004). Aristotle's logic. In: E. N. Zalta (ed.), *The Stanford Encyclopedia of Philosophy*. Center for the Study of Language and Information, Stanford University.

Song, X. M. and Montoya-Weiss, M. M. (1998). Critical development activities for really new versus incremental products. *Journal of Product Innovation Management*, **15**(2), 124–135.

Steinbock, D. (2003). Globalization of wireless value system: from geographic to strategic advantages. *Telecommunications Policy*, **27**(3-4), 207–235.

Steinbock, D. (2005). *The mobile revolution: the making of mobile services worldwide*. Kogan Page, London.

Steinmueller, W. E. (2000). Paths to convergence: the roles of popularisation, virtualisation and intermediation. In: E. Bohlin, K. Brodin, A. Lundgren, and B. Thorngren (eds.), *Convergence in Communications and Beyond*. North-Holland, Amsterdam.

Stieglitz, N. (2002). Industry dynamics and types of market convergence: the evolution of the handheld computers market in the 1990s and beyond. In: *DRUID Summer Conference on Industrial Dynamics of the New and Old Economy*, Copenhagen, Denmark. Danish Research Unit for Industrial Dynamics (DRUID).

Stieglitz, N. (2003). Digital dynamics and types of industry convergence: the evolution of the handheld computers market in the 1990s and beyond. In: J. F. Christensen and P. Maskell (eds.), *The Industrial Dynamics of the New Digital Economy*, pp. 179–208. Edwar Elgar, Cheltenham.

Stieglitz, N. (2004). *Strategie und Wettbewerb in konvergierenden Märkten*. VS Verlag für Sozialwissenschaften, Wiesbaden. Dissertation Universität Marburg 2003.

Stigler, G. (1951). The division of labor is limited by the extent of the market. *The Journal of Political Economy*, **59**(3), 185–193.

Streun, J. (2003). Marketing-Konzept für m-business-Lösungen in Schweizer Forschungsinstitutionen und artverwandten Organisationen. Unpublished report, ETH Zurich, Chair for Technology Management and Entrepreneurship.

Strübing, J. (2004). *Grounded Theory: Zur sozialtheoretischen und epistemologischen Fundierung des Verfahrens der empirisch begründeten Theoriebildung*. Qualitative Sozialforschung. VS Verlag für Sozialwissenschaften, Wiesbaden.

Suarez, F. F. and Utterback, J. M. (1995). Dominant designs and the survival of firms. *Strategic Management Journal*, **16**(6), 415–430.

SVEDA (2005). Nano-bio-info technology convergence. Report, Silicon Valley Economic Development Alliance (SVEDA).

Svendsen, G. T. and Fai, F. M. (2003). *Corporate Technological Competence and the Evolution of Technological Diversification*. Edward Elgar Publishing.

Tatsuno, K. (2006). Current trends in digital cameras and cameraphones. *Science & Technology Trends – Quarterly Review*, **18**(1), 35–44. Original Japanese version published in July 2005.

Teece, D. J. (1986). Profiting from technological innovation: Implications for integration, collaboration, licensing and public-policy. *Research Policy*, **15**(6), 285–305.

Teece, D. J., Rumelt, R., Dosi, G., and Winter, S. (1994). Understanding corporate coherence: Theory and evidence. *Journal Of Economic Behavior & Organization*, **23**(1), 1–30.

Teece, D. J., Pisano, G., and Shuen, A. (1997). Dynamic capabilities and strategic management. *Strategic Management Journal*, **18**(7), 509–533.

Tegart, G. (2002). Nanotechnology: the technology for the 21st century. Report, Asia-Pacific Economic Cooperation (APEC), Industrial Science and Technology Working Group.

Theilen, F. (2004). *Geschäftsmodellbasiertes Konvergenzmanagement auf dem Markt für mobile Financial Services*, vol. 36 of *Diskussionsreihe Bank & Börse*. Springer, Bank Verlag, Wien.

Thielmann, B. (2000). *Strategisches Innovations-Management in konvergierenden Märkten: Medien und Telekommunikationsunternehmen in Online-Diensten und im digitalen Fernsehen*. Deutscher Universitäts-Verlag, Wiesbaden.

Thomas, L. G. (1996). The two faces of competition: Dynamic resourcefulness and the hypercompetitive shift. *Organization Science*, **7**(3), 221–242.

Tidd, J., Bessant, J., and Pavitt, K. (2005). *Managing Innovation: Integrating Technological, Market and Organizational Change.* John Wiley & Sons, New York, NY, 3rd edition.

Tosi, H. L. and Slocum, J. W. (1984). Contingency theory: some suggested directions. *Journal of Management*, **10**(1), 9–26.

Trauffler, G. (2005). *Strategic management of discontinuous technologies and radical innovation.* Ph.D. thesis, ETH Zurich (Diss. ETH Nr. 16111).

Tripsas, M. (1997). Unraveling the process of creative destruction: complementary assets and incumbent survival in the typesetter industry. *Strategic Management Journal*, **18**(S1), 119–142.

Tushman, M. L. and Anderson, P. (1986). Technological discontinuities and organizational environments. *Administrative Science Quarterly*, **31**(3), 439–465.

Tushman, M. L. and O'Reilly, C. A. (1996). The ambidextrous organization: Managing evolutionary and revolutionary change. *California Management Review*, **38**(4), 8–30.

Tushman, M. L. and Romanelli, E. (1985). Organizational evolution: a metamorphosis model of convergence and reorientation. *Research in Organizational Behavior*, **7**, 171–222.

Tushman, M. L. and Rosenkopf, L. (1992). Organizational determinants of technological change: toward a sociology of technological evolution. *Research in Organizational Behavior*, **14**, 311–347.

Ulrich, H. (2001). Die Betriebswirtschaftslehre als anwendungsorientierte Sozialwissenschaft. In: H. Ulrich (ed.), *Gesammelte Schriften*, vol. 5, pp. 17–51. Verlag Paul Haupt, Bern.

Utterback, J. M. (1994). *Mastering the dynamics of innovation: how companies can seize opportunities in the face of technological change.* Harvard Business School Press, Boston, MA.

Utterback, J. M. and Abernathy, W. J. (1975). Dynamic model of process and product innovation. *Omega-International Journal of Management Science*, **3**(6), 639–656.

van de Ven, A. H. and Poole, M. S. (1989). Using paradox to build management and organization theories. *Academy of Management Review*, **14**(4), 562–578.

van de Ven, A. H. and Poole, M. S. (1995). Explaining development and change in organizations. *Academy of Management Review*, **20**(3), 510–540.

van Wegberg, M. (1995). Mergers and alliances in the multimedia market. Report, Maastricht Research School of Economics of Technology and Organization.

Vanhaverbeke, W. and Kirschbaum, R. (2003). Building new competencies for new business creation based on breakthrough technological innovations. Report, Eindhoven Centre for Innovation Studies, Working Paper 03.17.

Varis, J., Virolainen, V.-M., and Puumalainen, K. (2004). In search for complementarities: partnering of technology-intensive small firms. *International Journal of Production Economics*, **90**(1), 117–125.

Vesa, J. (2006). Regulatory framework and industry clockspeed. In: J. Müller and B. Preissl (eds.), *Governance of Communication Networks: Connecting Societies and Markets with IT*, pp. 79–90. Physica (Springer), Heidelberg.

Voelpel, S., Leibold, M., Tekie, E., and von Krogh, G. (2005). Escaping the red queen effect in competitive strategy: sense-testing business models. *European Management Journal*, **23**(1), 37–49.

Vogelsang, R. (2004). Development of a strategic market-entry framework for a high-density data storage system. Unpublished report, ETH Zurich, Chair for Technology Management and Entrepreneurship.

Vojak, B. A. and Chambers, F. A. (2004). Roadmapping disruptive technical threats and opportunities in complex, technology-based subsystems: The SAILS methodology. *Technological Forecasting and Social Change*, **71**(1-2), 121–139.

Volberda, H. W. (1996). Toward the flexible form: How to remain vital in hypercompetitive environments. *Organization Science*, **7**(4), 359–374.

von Hippel, E. (2005). *Democratizing innovation*. MIT Press, Cambridge, MA.

von Hippel, E. and von Krogh, G. (2003). Open source software and the 'private collective' innovation model: issue for organization science. *Organization Science*, **14**(2), 209–223.

von Hippel, E. and von Krogh, G. (2006). Free revealing and the private-collective model for innovation incentives. *R&D Management*, **36**(3), 295–306.

von Krogh, G. and von Hippel, E. (2006). The promise of research on open source software. *Management Science*, **52**(7), 975–983.

Walsh, S. T. (2004). Roadmapping a disruptive technology: A case study—the emerging microsystems and top-down nanosystems industry. *Technological Forecasting and Social Change*, **71**(1-2), 161–185.

Walsh, S. T., Kirchhoff, B. A., and Newbert, S. (2002). Differentiating market strategies for disruptive technologies. *IEEE Transactions on Engineering Management*, **49**(4), 341–351.

Walsh, S. T., Boylan, R. L., McDermott, C., and Paulson, A. (2005). The semiconductor silicon industry roadmap: Epochs driven by the dynamics between disruptive technologies and core competencies. *Technological Forecasting and Social Change*, **72**(2), 213–236.

Wasson, C. R. (1974). *Dynamic competitive strategy and product life cycles*. Challenge Books, St. Charles.

Watzlawick, P., Weakland, J. H., and Fisch, R. (1974). *Change: principles of problem formation and problem resolution*. W. W. Norton & Company, New York, NY.

Weigel, P. (2003). Generating a business case-based technology roadmap for the wireless network migration from GSM/GPRS towards WCDMA. Unpublished report, ETH Zurich, Chair for Technology Management and Entrepreneurship.

Wernerfelt, B. (1984). A resource-based view of the firm. *Strategic Management Journal*, **5**(2), 171–180.

Williams, L. (2006). Convergence creates skills problems. Computing, 22 May 2006.

Wind, Y., Mahajan, V., and Gunther, R. E. (2002). *Convergence marketing: strategies for reaching the new hybrid consumer*. Prentice Hall, Upper Saddle River, NJ.

Winter, S. G. (2003). Understanding dynamic capabilities. *Strategic Management Journal*, **24**(10), 991–995.

Wirth, U. V. (2004). Micropayment: internal and external approach. Unpublished report, ETH Zurich, Chair for Technology Management and Entrepreneurship.

Wirtz, B. W. (1999). Convergence processes, value constellations and integration strategies in the multimedia business. *International Journal on Media Management*, **1**(1), 14–22.

Wirtz, B. W. (2001). Reconfiguration of value chains in converging media and communications markets. *Long Range Planning*, **34**(4), 489–506.

Woodward, J. (1965). *Industrial organization: theory and practice*. Oxford University Press, Oxford.

Wunderlin, O. (2004). Value chain analysis in all-IP networks: the loss of voice business and additional revenue possibilities in all-IP networks. Unpublished report, ETH Zurich, Chair for Technology Management and Entrepreneurship.

Yang, D.-H., Kim, S., Nam, C., and Moon, J.-S. (2004). Fixed and mobile service convergence and reconfiguration of telecommunications value chains. *IEEE Wireless Communications*, **11**(5), 42–47.

Yin, R. (1994). *Case study research: design and methods*, vol. 2. Sage Publications, Thousand Oaks, CA.

Yin, R. (2003). *Applications of case study research*, vol. 34 of *Applied social research methods series*. Sage Publications, Thousand Oaks, CA.

Yoffie, D. B. (1996). Competing in the age of digital convergence. *California Management Review*, **38**(4), 31–53.

Yoffie, D. B. (1997). *Competing the age of digital convergence*. Harvard Business School Press, Boston, MA.

Yoon, E. and Lilien, G. L. (1985). New industrial-product performance: the effects of market characteristics and strategy. *Journal of Product Innovation Management*, **2**(3), 134–144.

Zerdick, A. (2000). *E-conomics: strategies for the digital marketplace*. Springer, Berlin.

Name Index

Subject Index

Printing: Krips bv, Meppel
Binding: Stürtz, Würzburg